Master Math: Pre-Calculus

By

Debra Anne Ross

Course Technology PTR
A part of Cengage Learning

Australia, Brazil, Japan, Korea, Mexico, Singapore, Spain, United Kingdom, United States

COURSE TECHNOLOGY
CENGAGE Learning™

Master Math: Pre-Calculus
Debra Anne Ross

Publisher and General Manager, Course Technology PTR: Stacy L. Hiquet

Associate Director of Marketing: Sarah Panella

Manager of Editorial Services: Heather Talbot

Marketing Manager: Jordan Casey

Senior Acquisitions Editor: Emi Smith

Interior Layout: Shawn Morningstar

Cover Designer: Jeff Cooper

Illustrations and Equations: Judith Littlefield

Indexer: Larry Sweazy

Proofreader: Jenny Davidson

Printed in Canada
2 3 4 5 6 7 11 10 09

Library of Congress Control Number: 2009924538
ISBN-13: 978-1-59863-981-0
ISBN-10: 1-59863-981-1

Course Technology, a part of Cengage Learning
20 Channel Center Street
Boston, MA 02210
USA

Cengage Learning is a leading provider of customized learning solutions with office locations around the globe, including Singapore, the United Kingdom, Australia, Mexico, Brazil, and Japan. Locate your local office at: **international.cengage.com/region.**

Cengage Learning products are represented in Canada by Nelson Education, Ltd.

For your lifelong learning solutions, visit **courseptr.com.**
Visit our corporate Web site at **cengage.com.**

Table of Contents

Acknowledgments

I sincerely thank Dr. Melanie McNeil, Professor of Chemical Engineering at San Jose State University, for reading this book for accuracy and for all her helpful comments. I am grateful to Dr. Channing Robertson, Professor of Chemical Engineering at Stanford University, for reviewing this book and, in general, for his sagacious guidance. I especially thank my mother, Maggie Ross, for reading this book and for her editorial help.

Without my wonderful agent, Sidney B. Kramer, and the staff of Mews Books, the *Master Math* series would not have been published. Thank you, Sidney! I am also thankful to Ron Fry and the staff of Career Press for their work in publishing and launching the original *Master Math* books as a successful series.

I am grateful to Emi Smith, Senior Acquisitions Editor, and Course Technology, a part of Cengage Learning, for invigorating the *Master Math* series and improving the presentation. I especially thank Shawn Morningstar's proficient work on layout and illustrations. I also very much appreciate Judith Littlefield's work on illustrations and equations. Much thanks to Jenny Davidson for proofreading, Jeff Cooper for cover design, Larry Sweazy for indexing, as well as Stacy L. Hiquet, Sarah Panella, Heather Talbot, and Jordan Casey.

Finally, I deeply appreciate my beautiful and brilliant husband, David A. Lawrence, who worked side-by-side with me as we meticulously edited text and figures.

About the Author

Debra Anne Ross Lawrence is the author of six books of the *Master Math* series: *Basic Math and Pre-Algebra*, *Algebra*, *Pre-Calculus*, *Calculus*, *Trigonometry*, and *Geometry*. She earned a double Bachelor of Arts degree in biology and chemistry with honors from the University of California at Santa Cruz and a Master of Science degree in chemical engineering from Stanford University.

Her research experience encompasses investigating the photosynthetic light reactions using a dye laser, studying the eye lens of diabetic patients, creating a computer simulation program of physiological responses to sensory and chemical disturbances, genetically engineering bacteria cells for over-expression of a protein, and designing and fabricating biological reactors for in-vivo study of microbial metabolism using nuclear magnetic resonance spectroscopy.

Debra was a member of a small team of scientists and engineers who developed and brought to market the first commercial biosensor system. She managed an engineering group responsible for scale-up of combinatorial synthesis for pharmaceutical development.

She also managed intellectual property for a scientific research and development company. Debra's work has been published in scientific journals and/or patented.

Debra is also the author of *The 3:00 PM Secret: Live Slim and Strong Live Your Dreams* and *The 3:00 PM Secret 10-Day Dream Diet*. She is the coauthor with her husband, David A. Lawrence, of *Arrows Through Time: A Time Travel Tale of Adventure, Courage, and Faith*. Debra is President of GlacierDog Publishing and Founder of GlacierDog.com. When Debra is not engaged in all-season mountaineering near her Alaska home, she is endeavoring to understand the incomprehensible workings of the universe.

Introduction

Pre-Calculus is the third book in the *Master Math* series. The series also includes *Basic Math and Pre-Algebra*, *Algebra*, *Geometry*, *Trigonometry*, and *Calculus*. The *Master Math* series presents the general principles of mathematics from grade school through college including arithmetic, algebra, geometry, trigonometry, pre-calculus, and introductory calculus.

Pre-Calculus is a comprehensive pre-calculus book that explains the subject matter in a way that makes sense to the reader. It begins with the most basic principles and progresses through more advanced topics to prepare a student for calculus. *Pre-Calculus* explains basic principles and operations of geometry, trigonometry, pre-calculus, and introductory calculus, provides step-by-step procedures and solutions and presents examples and applications.

Pre-Calculus is a reference book for middle school, high school, and college students that explains and clarifies the pre-calculus and calculus principles they are learning in school. It is also a comprehensive reference source for students currently learning pre-calculus and calculus. *Pre-Calculus* is invaluable for students, parents, tutors, and anyone needing a comprehensive pre-calculus reference source.

The information provided in each book and in the series as a whole is progressive in difficulty and builds on itself, which allows the reader to gain perspective on the connected nature of mathematics. The skills required to understand every topic presented are explained in an earlier chapter or book within the series. Each of the books contains a complete table of contents and a comprehensive index so that specific subjects, principles, and formulas can be easily found. The books are written in a simple style that facilitates understanding and easy referencing of sought-after principles, definitions, and explanations.

Pre-Calculus and the *Master Math* series are not replacements for textbooks but rather reference and teaching books providing explanations and perspective. The *Master Math* series would have been invaluable to me during my entire education from grade school through graduate school. There is no other source that provides the breadth and depth of the *Master Math* series in a single book or series.

Finally, mathematics is a language—the universal language. A person struggling with mathematics should approach it in the same fashion he or she would approach learning any other language. If someone moves to a foreign country, he or she does not expect to know the language automatically. It takes practice and contact with a language in order to master it. After a short time in the foreign country he or she would not say, "I do not know this language well yet. I must not have an aptitude for it." Yet many people have this attitude toward mathematics. If time is spent learning and practicing the principles, mathematics will become familiar and understandable. Don't give up.

Chapter 1

Geometry

1.1 Lines and Angles

• Different lines and angles have specific names according to their environment, position, size, measurement, etc. This section provides some common definitions for lines and angles.

• The shortest *distance between two points* is a straight line. The distance between points (x_1, y_1) and (x_2, y_2) is given by:

$$d = \sqrt{(x_2 - x_1)^2 + (y_2 - y_1)^2}$$

The coordinates of the midpoint between the two points are:

$$(x_1 + x_2)/2 \text{ and } (y_1 + y_2)/2$$

• A *line segment* is a section of a line between two points.

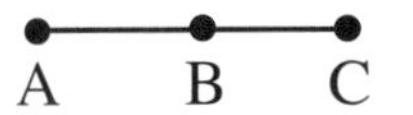

A B C

There are three line segments in the above line:

Segment AB, Segment BC, and Segment AC.

• A *ray* has an end point at one end and extends indefinitely in the other direction.

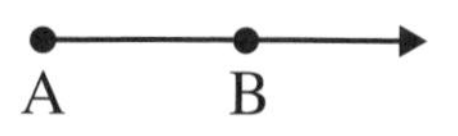

• If two straight lines meet or cross each other at a point, an *angle* is formed. The point where the lines meet is called the *vertex* of the angle and the sides are called the *rays* of the angle.

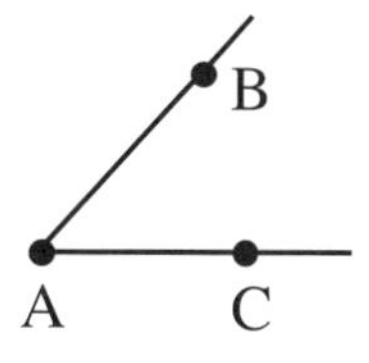

The vertex is at point A and the rays are AB and AC.

• The symbol for an angle is ∠. The above angle can be called ∠ BAC (where the middle letter names the vertex) or simply ∠ A.

• Angles are measured in *degrees* or *radians.* The symbol for degrees is °, and radian is often shortened to *rad.*

• A 180° angle is a straight line.

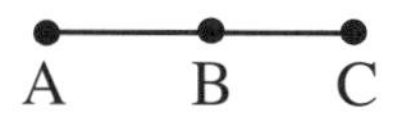

$\angle ABC = 180°$

• The symbol for a *right angle* or a 90° angle is a square drawn at the vertex.

• Angles smaller than 90° are called *acute angles.*

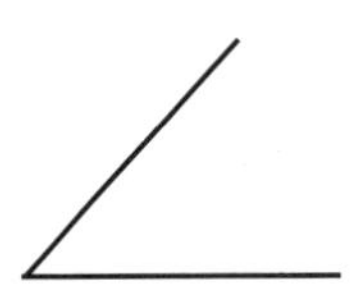

• Angles larger than 90° are called *obtuse angles.*

• If two angles have the same vertex and are adjacent to each other, they are called *adjacent angles.*

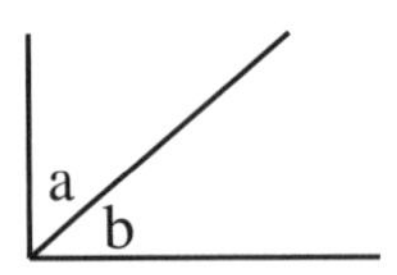

$\angle$ a and $\angle$ b are adjacent angles.

• If the sum of any two angles equals 180°, the two angles are called *supplementary angles.* The following examples are of angles that are adjacent and supplementary. Note that supplementary angles do not have to be adjacent.

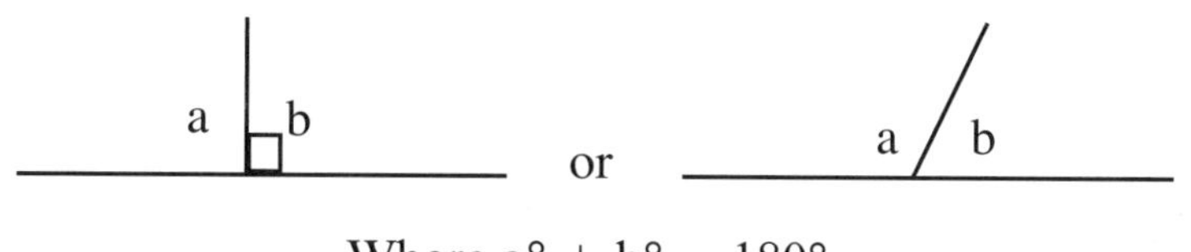

Where $a° + b° = 180°$.

• If the sum of any two angles equals 90°, the two angles are called *complementary angles.* For example, if ∠ABC is a right angle, then angles ∠ABD and ∠DBC are adjacent and complementary angles. Note that complementary angles do not have to be adjacent.

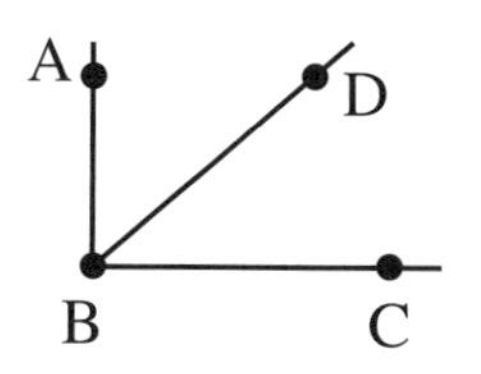

• If two lines intersect each other, there are four angles formed.

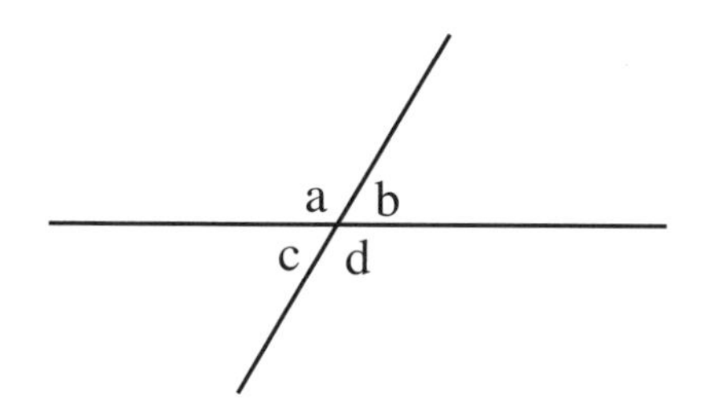

The sum of the four angles is 360°.

$$a° + b° + c° + d° = 360°$$

The sum of the adjacent angles is 180°.

$a° + b° = 180°$

$c° + d° = 180°$

$a° + c° = 180°$

$b° + d° = 180°$

The angles opposite to each other are called *vertical angles* and they are equivalent.

$a° = d°$

$b° = c°$

• If two lines are perpendicular to each other, four *right angles*, each measuring 90°, are formed.

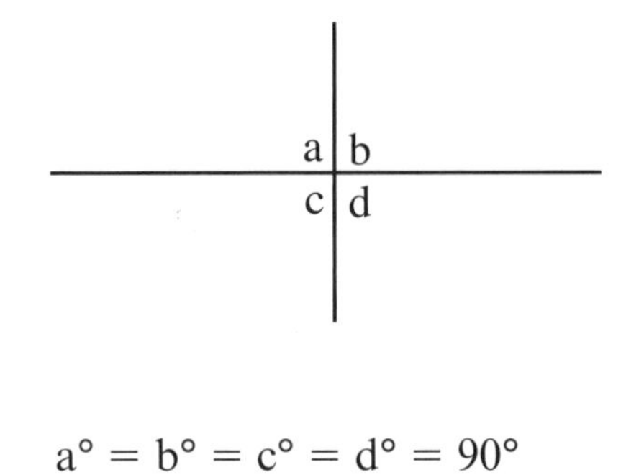

$a° = b° = c° = d° = 90°$

$a° + b° + c° + d° = 360°$

- A *transversal* is a line that intersects two other lines.

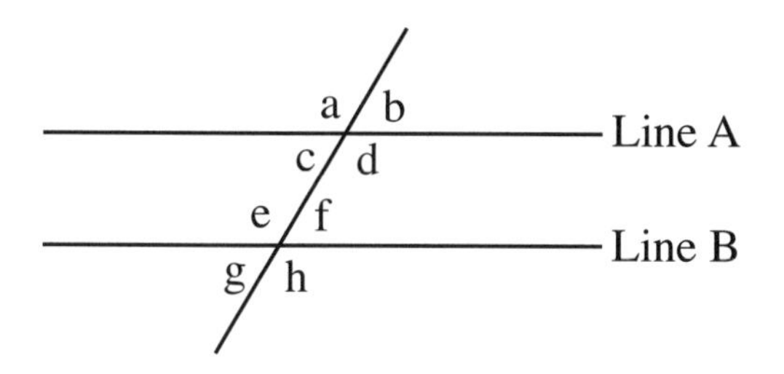

a, b, c, d, e, f, g, and h represent angles.

If lines A and B are parallel to each other, the following is true:

Angles a = d = e = h

Angles b = c = f = g

Angle c + Angle e = 180°

Angle d + Angle f = 180°

- As described later in the circle section, there are:

360° in a circle.

180° in a semi-circle.

90° in a quarter-circle.

45° in an eighth-circle.

1.2 Polygons

• A *polygon* is a closed planar figure that is formed by three or more line segments that all meet at their end points, and there are no end points that are not met by another end point. The line segments that make up a polygon only intersect at their end points.

• Examples of polygons are:

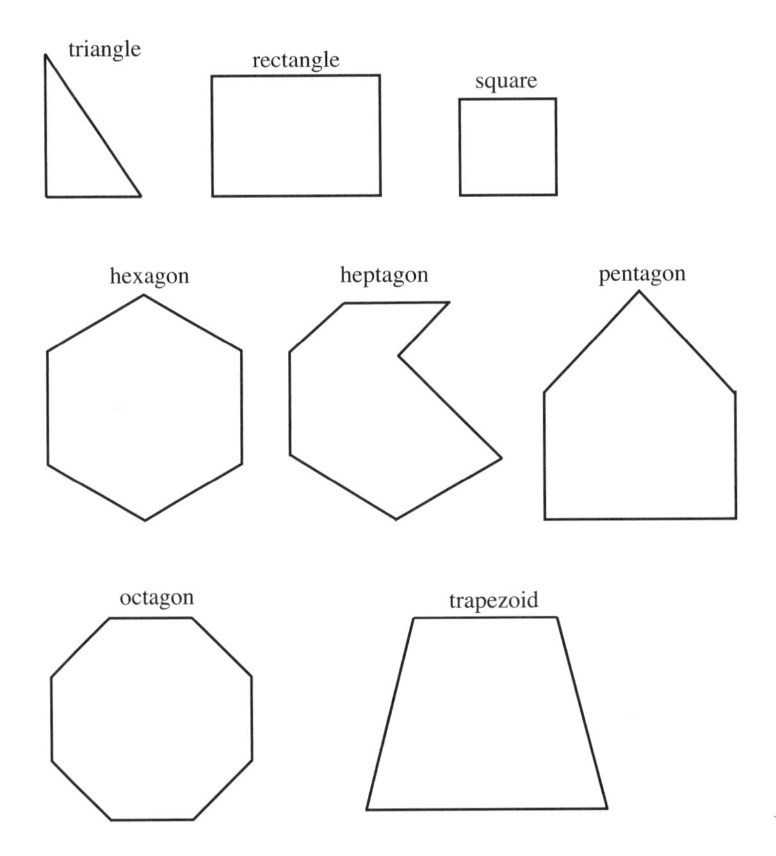

Note that rectangles, squares, and trapezoids are also quadrilaterals.

• Polygons are named according to the number of sides they have. For example:

Number of Sides and Name

3 = Triangle
4 = Quadrilateral
5 = Pentagon
6 = Hexagon
7 = Heptagon
8 = Octagon
9 = Nonagon
10 = Decagon

• In a polygon, the number of sides equals the number of angles.

• Polygons are often referred to by labels or letters written at their angles. For example, the following quadrilateral can be referred to as ABDC:

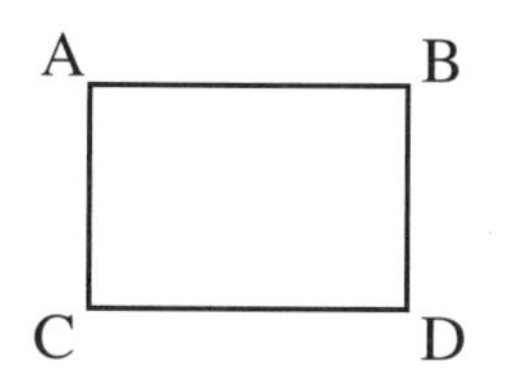

• If the length of the sides are equal and the angle measurements are equal, the polygon is called a *regular polygon.* A square is a *regular quadrilateral.*

• If two polygons have the same size and shape, they are called *congruent polygons.*

• If two polygons have the same shape such that their angle measurements are equal and their sides are proportional, however one is larger than the other, they are called *similar polygons.*

• The following equation gives the sum of all the angle measurements in a polygon:

$(n - 2)180° =$ Sum of all angles in n-gon

$n =$ the number of angles (or sides) in the polygon.

• For example, if a quadrilateral has four sides and four angles, what is the sum of all angle measurements?

$(n - 2)180° =$ Sum of all angles in n-gon

$n = 4$

$(4 - 2)180° = (2)180° = 360°$

Therefore, the sum of the angles in a quadrilateral is 360°.

1.3 Triangles

• *Triangles* are three-sided polygons. The symbol for a triangle is Δ.

• The sum of the angles of a triangle is:

$$(n - 2)180°$$

For a triangle, $n = 3$,

$$(3 - 2)180° = (1)180° = 180°$$

Therefore, the sum of the angles of every triangle is 180°.

• If the value of two angles in a triangle is known, the third angle can be calculated by subtracting the sum of the two known angles from 180°.

• The length of one side of a triangle is always less than the sum of the other two sides.

• In a triangle, the largest side is opposite the largest angle, the smallest side is opposite the smallest angle, and the middle-length side is opposite the middle-size angle.

- Examples of triangles are:

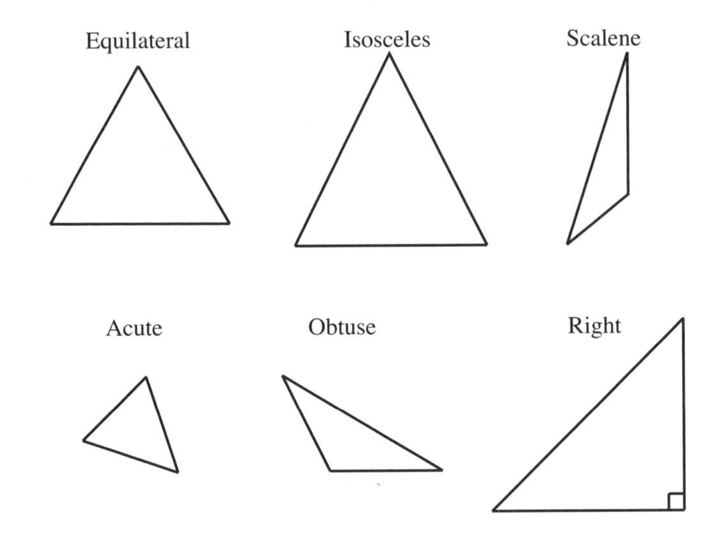

- In an *equilateral triangle,* all three sides have equal lengths and all three angles have equal measurements of 60°. The sum of the angles in an equilateral triangle is: $60° + 60° + 60° = 180°$.

- In an *isosceles triangle*, two sides have equal lengths and the angles opposite the two equal sides have equal measurements.

- In a *scalene triangle,* all three sides have different lengths and all three angles have different measurements.

- If one of the angles in the triangle is a right angle (90°), the triangle is called a *right triangle.*

- If all of the angles in a triangle are smaller than 90°, the triangle is called an *acute triangle.*

- If one of the angles in the triangle is larger than 90°, the triangle is called an *obtuse triangle.*

- In a *right triangle*, the side opposite to the right angle is called the *hypotenuse* and the two sides that meet to form the right angle are called *legs.* The hypotenuse is always the longest side.

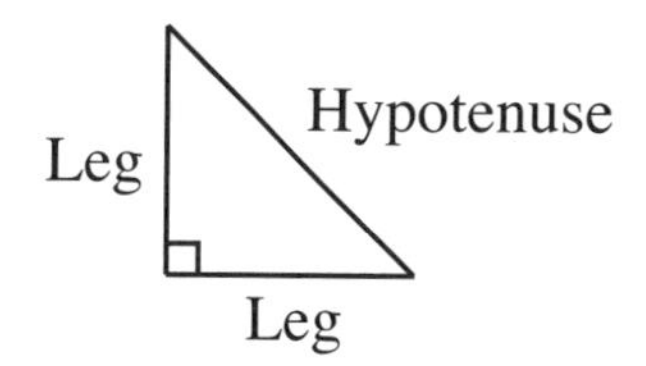

- In a right triangle, the square of the length of the hypotenuse is equal to the sum of the squares of the lengths of the legs.

$$(\text{Leg})^2 + (\text{Leg})^2 = (\text{Hypotenuse})^2$$

This is called the *Pythagorean Theorem* and it only applies to right triangles.

If the lengths of the legs are x and y and the length of the hypotenuse is z, the Pythagorean Theorem can be written:

$$x^2 + y^2 = z^2$$

• Two noteworthy right triangles are the 30°-60°-90° and the 45°-45°-90° (right isosceles triangle).

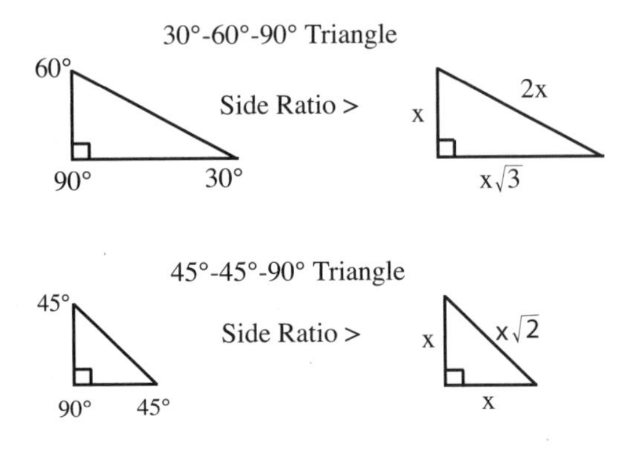

• Other noteworthy right triangles are called *triplets*. The most common triplets are 3:4:5, 5:12:13, and 7:24:25.

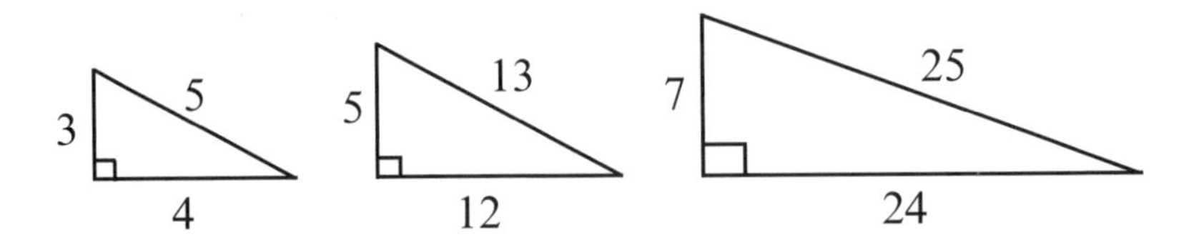

Note that any multiple of the ratios of these triangles is also a triplet, such as 6:8:10, 10:24:26, etc.

• If two corresponding angles of two triangles are equal, then the third angles are also equal.

• If two corresponding sides of two right triangles are equal, the third corresponding sides are also equal and the triangles are *congruent*. This can be proven using the Pythagorean Theorem.

- Two triangles are called *congruent triangles* if:

1. All three corresponding sides are equal; this is called *side-side-side.*

2. Two corresponding sides with their vertex angles are equal; this is called *side-angle-side.*

3. Two corresponding angles with the side in between are equal; this is called *angle-side-angle.*

side-side-side

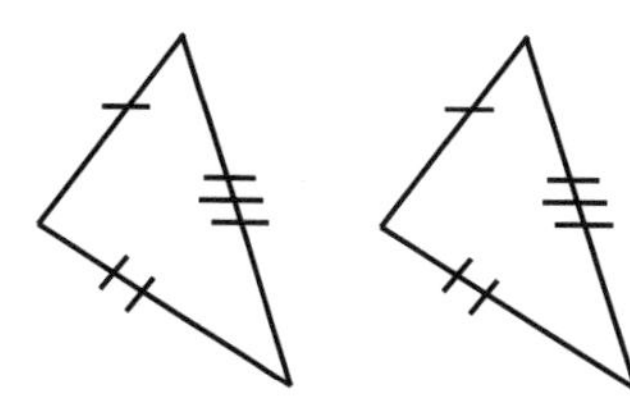

side-angle-side

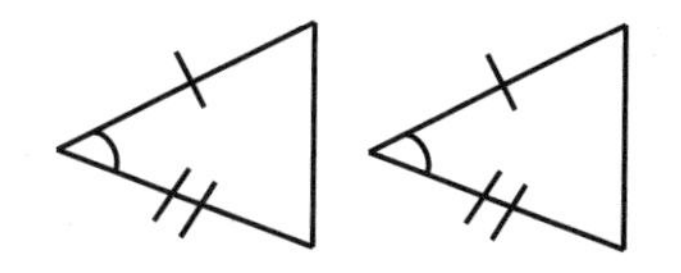

angle-side-angle

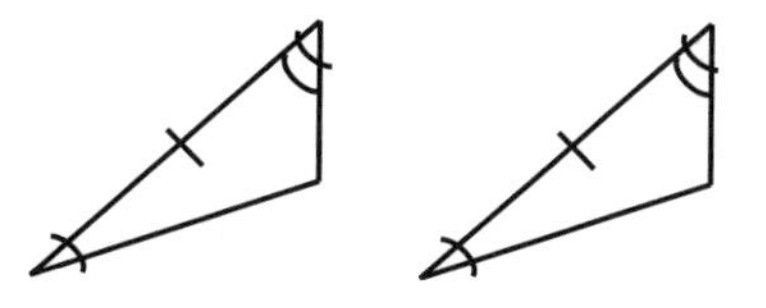

(Slash marks indicate same length.)

- If all three pairs of corresponding angles in two triangles are equal to each other, the two triangles are called *similar triangles*. Two similar triangles can be created by drawing a line parallel to one of the sides of a triangle.

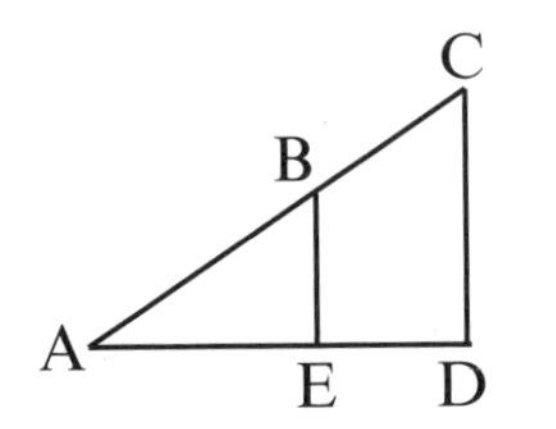

ΔADC is similar to ΔAEB because three corresponding angles are equal.

1.4 Quadrilaterals (Four-Sided Polygons)

- In this section, various types of quadrilaterals including parallelograms, rhombuses, rectangles, squares, and trapezoids are described.

- *Quadrilaterals* are four-sided polygons.

- The sum of the angles in a quadrilateral is:

$$(n - 2)180° = (4 - 2)180° = (2)180° = 360°$$

The sum of the angles in all quadrilaterals is 360°.

• A *parallelogram* is a quadrilateral with both of the two opposite sides parallel to each other.

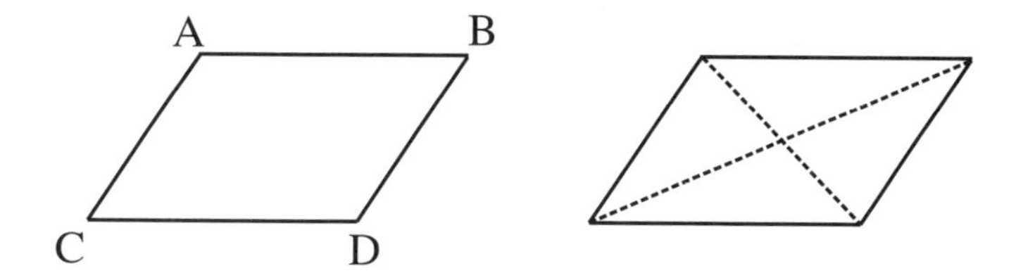

AB is parallel to CD. AC is parallel to BD. Both pairs of opposite sides are the same length. Both pairs of opposite angles are the same size. If diagonal lines (shown dashed above) are drawn to either pair of opposite angles the resulting triangles are congruent.

• A *rhombus* is a parallelogram with all four sides of equal length.

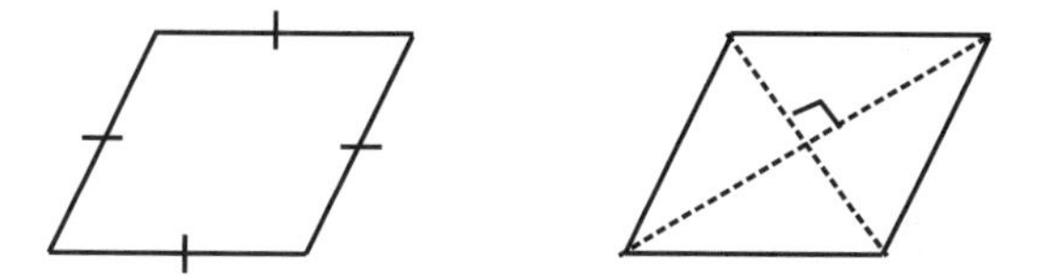

(Slash marks indicate same length)

Opposite angles of a rhombus are equal to each other. Diagonal lines (shown dashed above) drawn in a rhombus are perpendicular to each other.

- A *rectangle* is a parallelogram with all four angles having equal measurements of 90°.

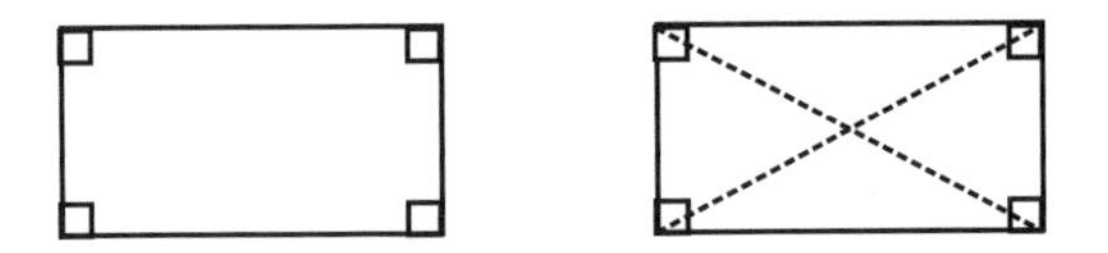

Opposite sides of a rectangle are also equal to each other. Diagonal lines (shown dashed above) drawn in a rectangle are the same length. The length of the diagonal lines can be determined using the Pythagorean Theorem if the side lengths of the rectangle are known.

$$(\text{long-side length})^2 + (\text{short-side length})^2 = (\text{diagonal length})^2$$

- A *square* is a parallelogram in which the four angles and four sides are equal.

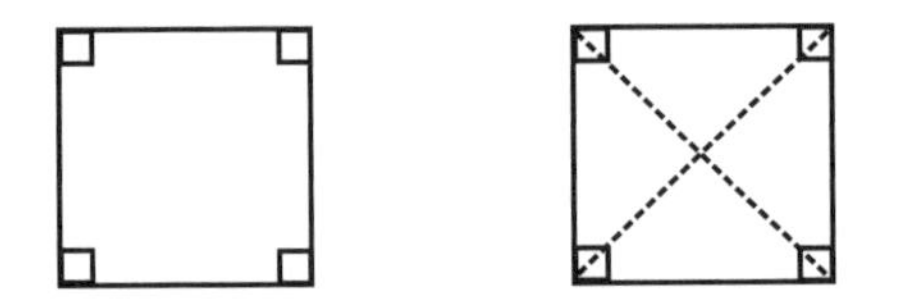

The angles will each measure 90°. Diagonal lines (shown dashed above) drawn in a square are the same length. The length of the diagonal lines can be determined using the Pythagorean Theorem if the side lengths of the square are known.

$$(\text{side length})^2 + (\text{side length})^2 = (\text{diagonal length})^2$$

For example, if the side length is x and the diagonal length is d:

$$x^2 + x^2 = d^2$$

$$d^2 = 2x^2$$

$$d = \sqrt{2x^2}$$

$$d = x\sqrt{2}$$

• A *trapezoid* is a quadrilateral having only one pair of opposite sides parallel to each other.

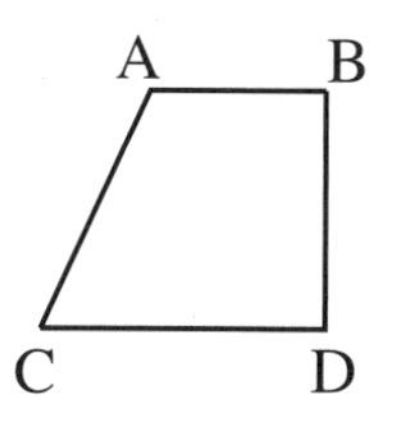

AB is parallel to CD. AC is not parallel to BD. AB and CD are called *bases*. AC and BD are called *legs*.

1.5 Circles

• In this section, the following terms are defined: circle, radius, diameter, chord, Pi, degree, radian, tangent, concentric, central angle, inscribed angle, and arc.

• A *circle* is a planar shape consisting of a closed curve in which each point on the curve is the same distance from the center of the circle.

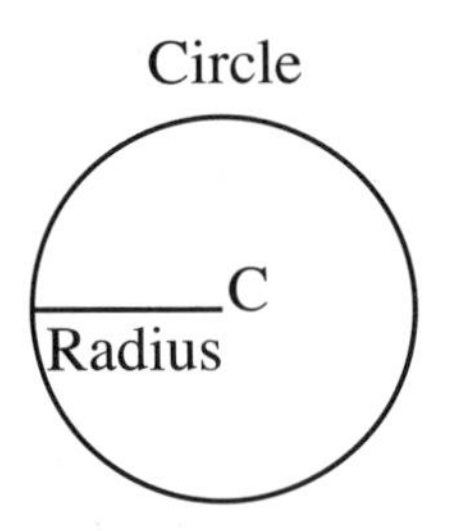

C denotes the center of the circle.

• The *radius* of a circle is the distance between the center and any point on the circle. All radii drawn for a given circle have the same length.

• A line segment drawn through the center point with its end points on the circle is called the *diameter* of the circle. The diameter is twice the radius.

$$(2)(\text{Radius}) = \text{Diameter}$$

A diameter line divides a circle into two equal semicircles.

• Any line segment whose ends are on the circle is called a *chord* (including the diameter line segment).

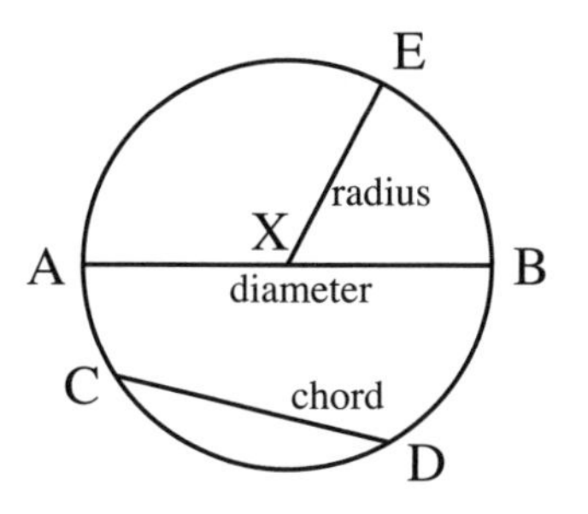

X depicts the point at the center of the circle. AB is the diameter chord. CD is a chord. XE, AX, and XB are radii.

• Pi, or π, defines the ratio between the circumference and the diameter of a circle. More specifically, Pi is equivalent to the circumference divided by the diameter of a circle.

• The value of Pi is approximately 3.141592654.

• The degrees of a circle are:

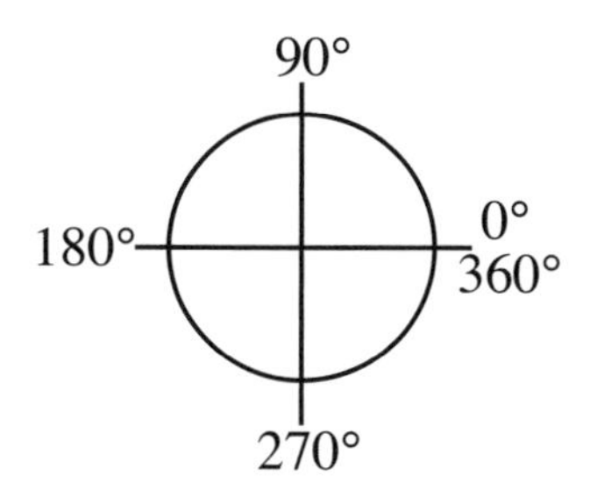

2π radians = 360 degrees.

1 radian = $360°/2\pi = 180°/\pi$.

1 degree = 2π radians/ $360° = \pi$ radians/$180°$.

Because there are 360° in a circle: 1° = 1/360.

1 *Minute* is defined as (') = (1/60) of 1°.

1 *Second* is defined as ('') = (1/60) of 1 Minute.

• A circle always measures 360° around, equivalent to 2π radians. Half of the circle measures 180°, which is equivalent to π radians. A quarter of a circle measures 90°, which is equivalent to $\pi/2$ radians.

• A *tangent line* passes through only one point on a circle. If a radius line segment is drawn from the center of the circle to the point of tangency, the tangent line and the radius line segment are perpendicular to each other.

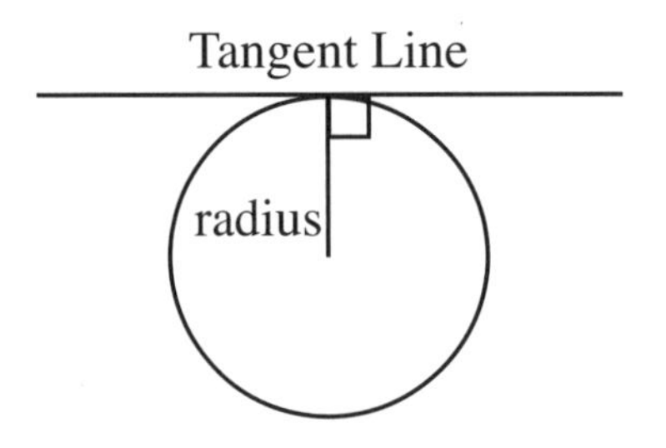

• If two or more circles have the same center point they are called *concentric circles.*

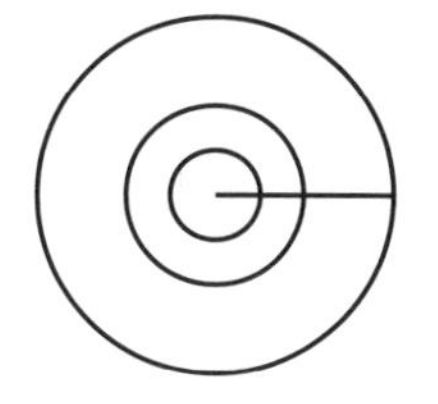

• If angles or polygons are drawn inside circles, any angle whose vertex is at the circle's center point is called a *central angle.*

• If angles or polygons are drawn inside circles, any angle whose vertex is on the circle is called an *inscribed angle.*

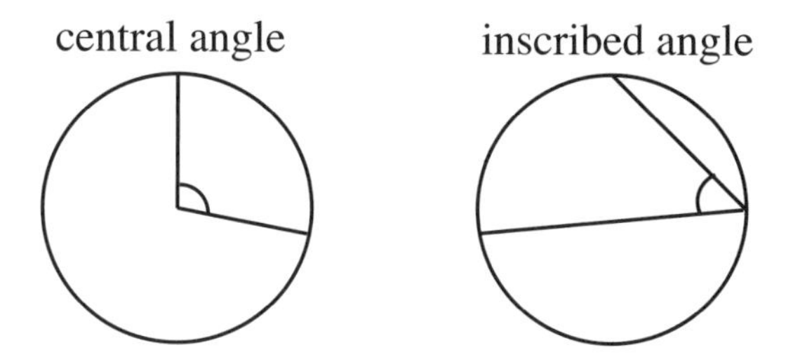

• A section of a circle defined by two or more points is called an *arc.* An arc can be measured in degrees or radians. (There are 2π radians in 360° and π radians in 180°.)

Arc length = (radius)×(central angle measure in radians)

or

Arc length = $(n°/360°)\times(\pi d)$

Where n° is the central angle.

For example, if r = 10, d = 20, n = 90° = π/2 radians,

Arc length = (r)(n) = (10)(π/2) = 15.7

or

Arc length = (n°/360°)(πd) = (90°/360°)(20π) = 15.7

- The measure of a central angle is equal to the measure of the arc it intercepts.

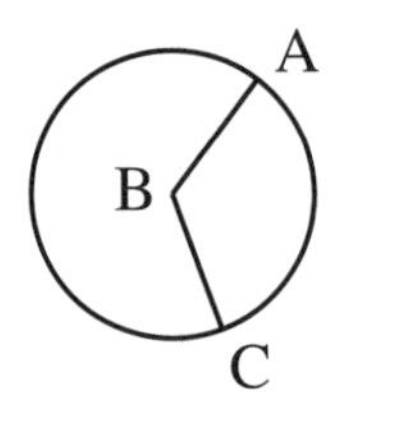

∠ABC = Arc AC

Note the following relationship for central angles:

$$\frac{\text{measure of angle B}}{360} = \frac{\text{length of arc AC}}{\text{circumference}} = \frac{\text{area of sector BAC}}{\text{area of circle}}$$

- A central angle subtending an arc equal in length to the radius of a circle is defined as a radian. One radian is approximately 57.29578 degrees. (Subtend means to extend under or be opposite to.)

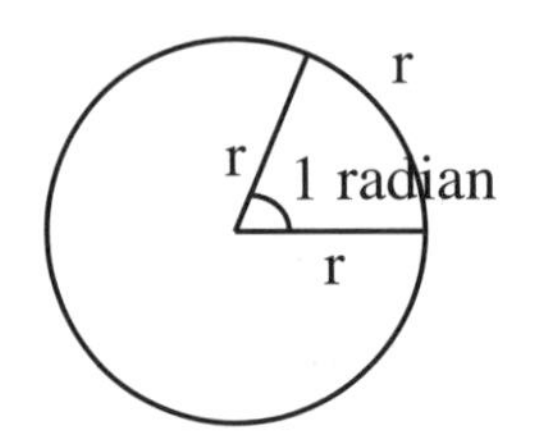

• The measure of an inscribed angle is equal in measure to half of the arc it intercepts.

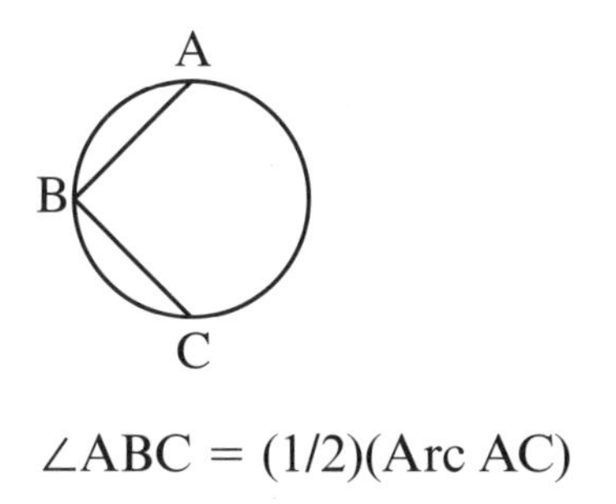

$\angle ABC = (1/2)(\text{Arc } AC)$

• If an inscribed angle has its rays ending at the end points of a diameter chord, the vertex of the angle will be a right angle (90°), (which is one-half of the 180° measurement of the arc).

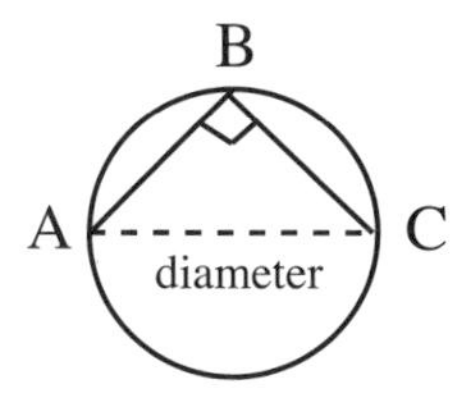

$\angle ABC = 90° = (1/2)(\text{Arc } AC) = (1/2)180°.$

• Inscribed angles with the same endpoints defined by the same arc, have the same measure.

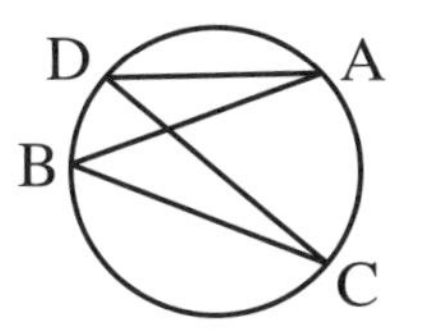

Angle ADC has the same measure as angle ABC.

• An inscribed angle is equal to one-half of the central angle formed from the endpoints of the same arc.

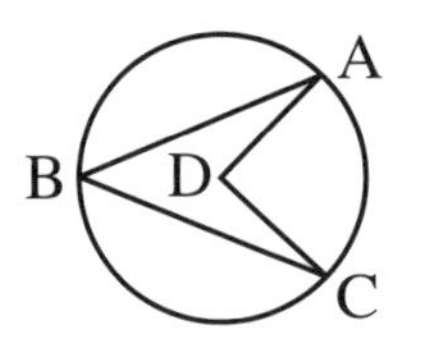

Angle ABC is an inscribed angle with endpoints on arc AC. Angle ADC is a central angle with endpoints on arc AC. Because they are formed by the same arc, angle ABC is one-half the measure of angle ADC.

1.6 Perimeter and Area of Planar Two-Dimensional Shapes

• Measuring perimeter and area of two-dimensional objects including triangles, squares, rectangles, regular hexagons, circles, parallelograms, and trapezoids are described in this section.

• The *perimeter* of polygons and planar figures such as circles is the sum of the lengths of its sides or the distance around something. The units for perimeter are always singular because of the one-dimension described. Remember to convert all measurements to the same units before adding.

- The following are examples of perimeter measurements.

Triangle

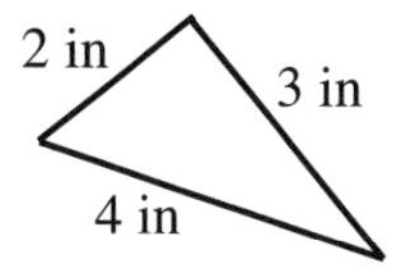

Perimeter of a triangle is the sum of the sides:

2 inches + 3 inches + 4 inches = 9 inches

Square

2 in

Perimeter of a square is the sum of the sides:

2 inches + 2 inches + 2 inches + 2 inches

= 4(2 inches) = 8 inches

Rectangle

2 in

4 in

Perimeter of a rectangle is the sum of the sides:

2 inches + 2 inches + 4 inches + 4 inches

= 2(2 inches) + 2(4 inches) = 12 inches

Regular Hexagon

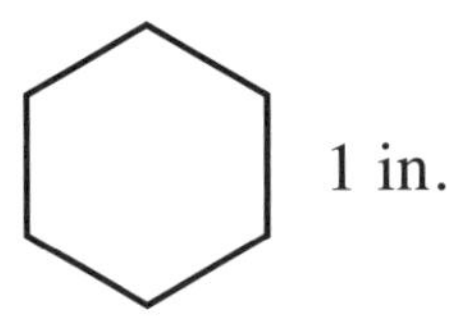

Perimeter of a hexagon is the sum of all the sides:

1 inch + 1 inch + 1 inch + 1 inch + 1 inch + 1 inch

= 6(1 inch) = 6 inches

Circle (The perimeter of a circle is called the *circumference*.)

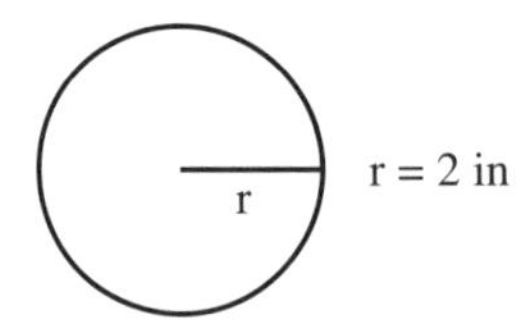

Perimeter of a circle = circumference = $2\pi r = \pi d$

$= 2\pi 2$ inches = 4π inches $\approx$ 12.56 inches

Where, r = radius, d = diameter, and $\pi \approx 3.14$.

- The *area* of polygons and planar figures, such as circles, is a measure of the planar dimensional space that the figure occupies. The units for area are always squared because of the two dimensions described, $(x)(x) = x^2$. Remember to convert all measurements to the same units before calculating.

• The following are examples of area measurements.

Triangle

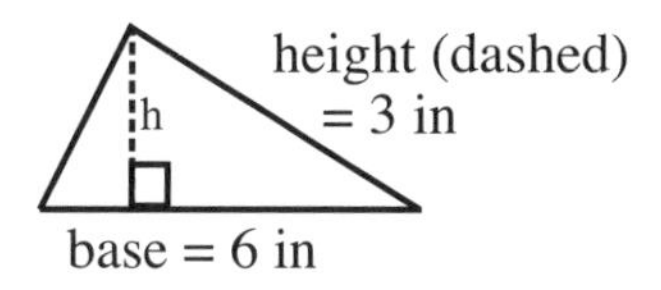

Area of a triangle = (1/2)(base)(height) =
(1/2)(6 inches)(3 inches) = 9 inches^2 or 9 square inches

To obtain the height, draw a line perpendicular from the base to the opposite angle.

Square

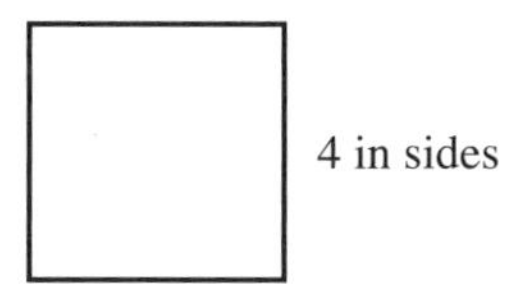

Area of a square = (side length)^2
= (4 inches)^2 = 16 inches^2 or 16 square inches

Rectangle

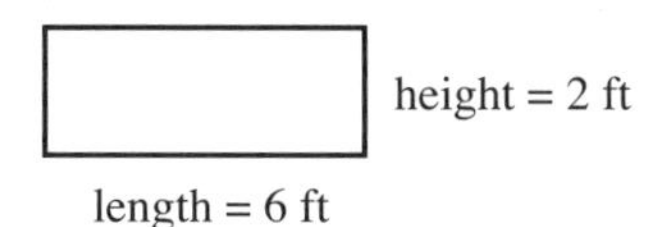

Area of a rectangle = (length)(height)
= (6 feet)(2 feet) = 12 feet^2 or 12 square feet

Parallelogram

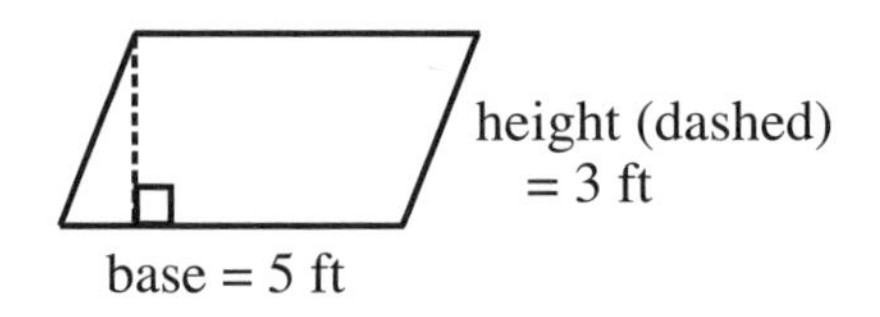

Area of a parallelogram = (base)(height)
= (5 feet)(3 feet) = 15 feet^2 or 15 square feet

To obtain the height, draw a line perpendicular from the base to the opposite angle.

Trapezoid

base = 4 ft

height (dashed)
= 5 ft

base = 8 ft

Area of a trapezoid = (average of bases)(height)
= (4 feet + 8 feet)/(2) × (5 feet)
= (6 feet)(5 feet) = 30 feet^2 or 30 square feet

To obtain the height, draw a line perpendicular from the base to the opposite angle.

• To find the area of polygons that are not triangles, squares, rectangles, parallelograms, or trapezoids, find the area of sections of the polygon that form one of these figures, then add the areas of the sections.

Circle

r

r = 2 in

Area of a circle $= \pi r^2 = \pi(d/2)^2$

$= \pi(2 \text{ inches})^2 = \pi 4 \text{ inches}^2 \approx 12.56$ square inches

Where r = radius, d = diameter, and $\pi \approx 3.14$.

The area of a *sector* of a circle is a fraction of the area of the whole circle. For a circle with its central angle = ABC:

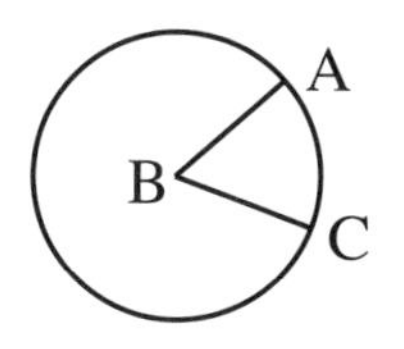

The following relationship is true:

$$\frac{\text{measure of angle B}}{360} = \frac{\text{length of arc AC}}{\text{circumference}} = \frac{\text{area of sector BAC}}{\text{area of circle}}$$

If B = 60°, then the central angle is 60/360, or 1/6 of the circle. Also, the length of arc AC is 1/6 of the circumference of the circle. Finally, the area of sector BAC is 1/6 of the area of the circle.

1.7 Volume and Surface Area of Three-Dimensional Objects

• In this section, measuring volume, surface area, and main diagonal of three-dimensional objects including cubes, rectangular solids, cylinders, spheres, ellipsoids, cones, and pyramids are described.

• *Volume* is a measure of the three-dimensional space that an object occupies. The units for volume are always cubed because of the three dimensions described $(x)(x)(x) = x^3$. Remember to convert all measurements to the same units before calculating.

• The *surface area* of three-dimensional objects such as cubes, rectangular solids, cylinders, and spheres is a sum of the areas of the surfaces. The units for surface area are always squared because of the two dimensions described, $(x)(x) = x^2$. Remember to convert all measurements to the same units before calculating.

• The following are examples of volume, surface area, and main diagonal measurements.

Cubes have six surfaces that are each squares and have the same measurements.

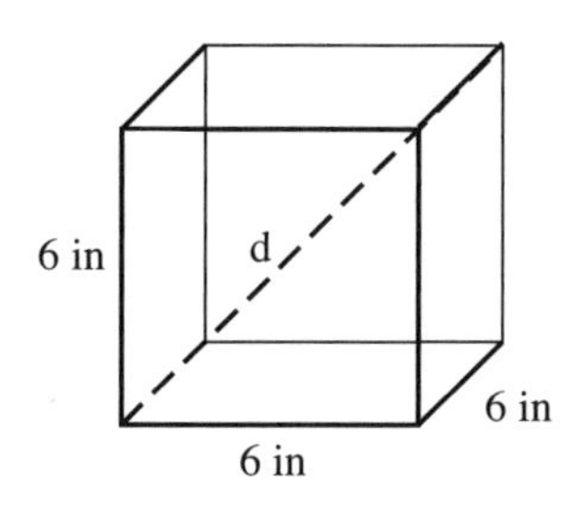

Volume of a cube = $(\text{edge})^3$

$= (6 \text{ inches})^3 = (6 \text{ inches})(6 \text{ inches})(6 \text{ inches})$

$= 216 \text{ inches}^3$ or 216 cubic inches

Surface area of a cube = (6 sides)(area of each side)

$= (6 \text{ sides})((6 \text{ inches})(6 \text{ inches})) = (6 \text{ sides})(36 \text{ inches}^2)$

$= 216 \text{ inches}^2$ or 216 square inches

The *main diagonal* of a cube is given by:

$s^2 + s^2 + s^2 = 3s^2 = d^2$

Where s = the length of the edge.

Rectangular solids have six rectangular surfaces with three pairs of opposite surfaces that have the same measurements.

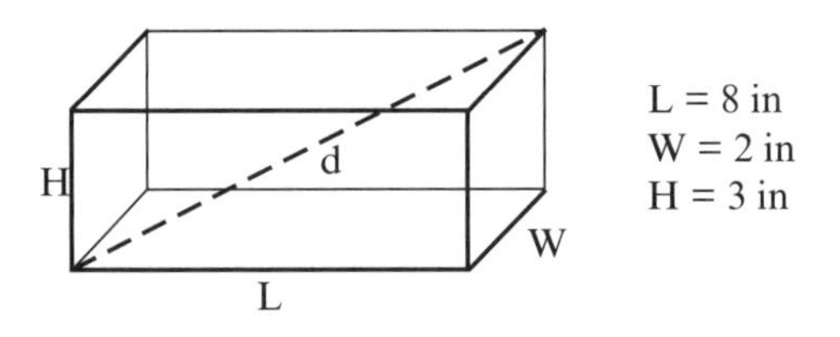

Volume of a rectangular solid

= (length)(width)(height)

= (8 in.)(2 in.)(3 in.) = 48 inches^3 or 48 cubic inches

Surface area of a rectangular solid

= (2)(length)(width) + (2)(length)(height) + (2)(width)(height)

= (2)(8 in.)(2 in.) + (2)(8 in.)(3 in.) + (2)(2 in.)(3 in.)

= (2)(16 in.^{2}) + (2)(24 in.^{2}) + (2)(6 in.^{2})

= (32 in.^{2}) + (48 in.^{2}) + (12 in.^{2}) = 92 in.^{2} or 92 square inches

Main diagonal of a rectangular solid is given by:

$$l^2 + w^2 + h^2 = d^2$$

Cylinders or circular solids are three-dimensional objects that have two identical circles connected by a tube.

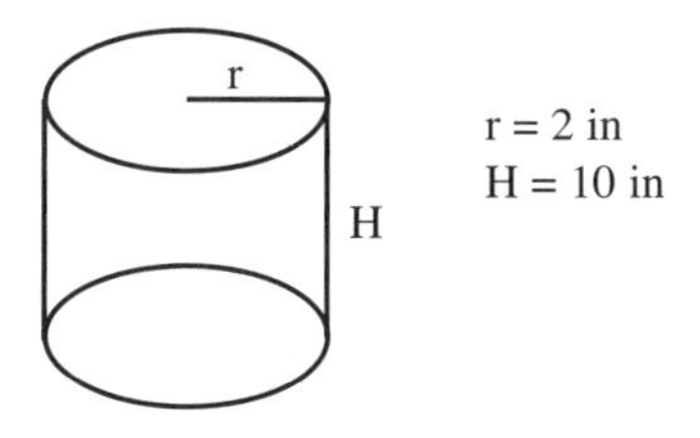

Volume of a cylinder or circular solid

= (area of circle)(height) = $(\pi r^2)(h)$

= $(\pi)(2 \text{ inches})^2(10 \text{ inches}) = (3.14)(4 \text{ inches}^2)(10 \text{ inches})$

= 125.6 inches^3 or 125.6 cubic inches

Surface area of a cylinder or circular solid

= (area of both circles) + (area of tube) = $2\pi r^2 + 2\pi rh$

(where $2\pi r$ = circumference)

= $(2)(\pi)(2 \text{ inches})^2 + (2)(\pi)(2 \text{ inches})(10 \text{ inches})$

= $(2)(3.14)(4 \text{ inches}^2) + (2)(3.14)(20 \text{ inches}^2)$

= $25.12 \text{ in.}^2 + 125.6 \text{ in.}^2 = 150.72 \text{ in.}^2$ or 150.72 square inches

- Note: Pi is approximately equal to 3.1415926535897932384626433832795... and often abbreviated 3.14.

Spheres or spherical solids are three-dimensional objects consisting of points that are all the same distance from the center.

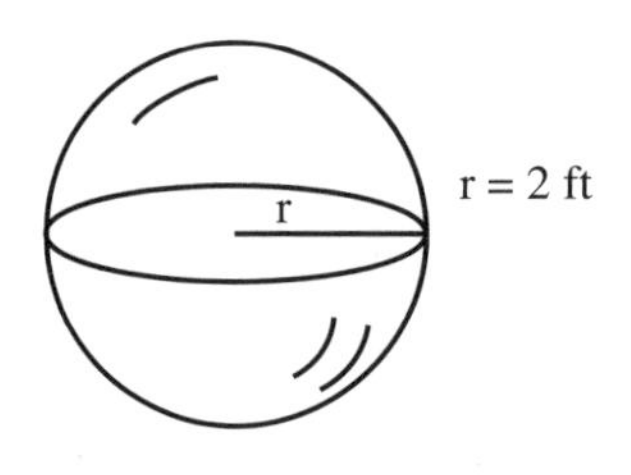

Volume of a sphere $= (4/3)\pi r^3$

$= (4/3)(\pi)(2 \text{ feet})^3 = (4/3)(3.14)(8 \text{ feet}^3)$

$= 33.49 \text{ feet}^3$ or 33.49 cubic feet

Surface area of a sphere $= 4\pi r^2$

$= 4\pi(2)^2 = 4(3.14)(2 \text{ ft.})^2 = 50.24 \text{ feet}^2$ or 50.24 square feet

The equation for the surface of a sphere drawn in the center of a coordinate system is:

$x^2 + y^2 + z^2 = r^2$

Ellipsoids are oval-shaped, three-dimensional objects.

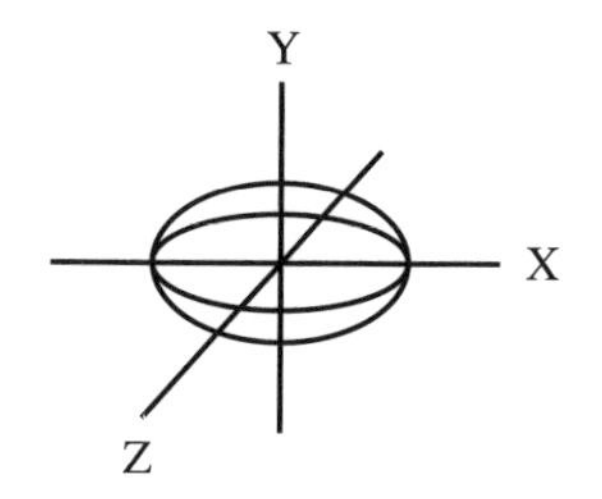

The equation for the surface of an ellipsoid drawn in the center of a coordinate system is:

$$\frac{x^2}{a^2} + \frac{y^2}{b^2} + \frac{z^2}{c^2} = 1$$

Cones are three-dimensional objects that have a circle connected to a point. The depth of a cone forms a triangular solid.

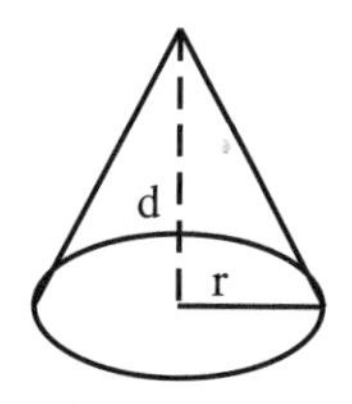

Volume of a cone

= (1/3)(area of circle)(height) = $(1/3)\pi r^2 d$

Note that the volume of a cone is one-third the volume of a cylinder of the same radius and height.

Pyramids are three-dimensional objects that have a square, rectangle, triangle, or other polygonal base connected to a point.

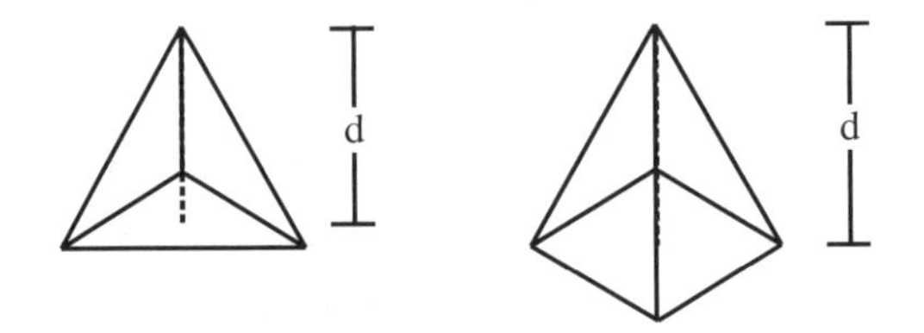

Volume of a pyramid

= (1/3)(area of base)(height) = (1/3)(area of base)(d)

Remember, the area of a triangle is (1/2)(base)(height), the area of a square is $(\text{side})^2$, and the area of a rectangle is (length)(height).

1.8 Vectors

- *Scalars* represent quantities that can be described by one number (either positive, negative, or zero). However, *vectors* represent quantities that must be described by a numerical value and a direction. Examples of scalars are temperature and mass. Examples of vectors are velocity and force.

- *Vectors* are represented as a line with an arrow pointing in one direction. For example, the following depicts a vector:

$$\longrightarrow$$

• Vectors that point in the same direction and have the same length are equivalent.

• To add vectors, position the vectors so that the beginning (initial point) of the second vector is at the end (final point) of the first vector. The sum of the two vectors will be a third vector with its initial point at the initial point of the first vector and its final point at the final point of the second vector.

• Example: Add vectors **a** and **b** in the following two examples.

In the first example, put the initial point of **b** at the final point of **a**. The sum is the vector joining the initial point of **a** to the final point of **b**, or vector **c**.

In the second example, the initial point of **b** is already at the final point of **a**. The sum is the vector joining the initial point of **a** to the final point of **b,** or vector **c**.

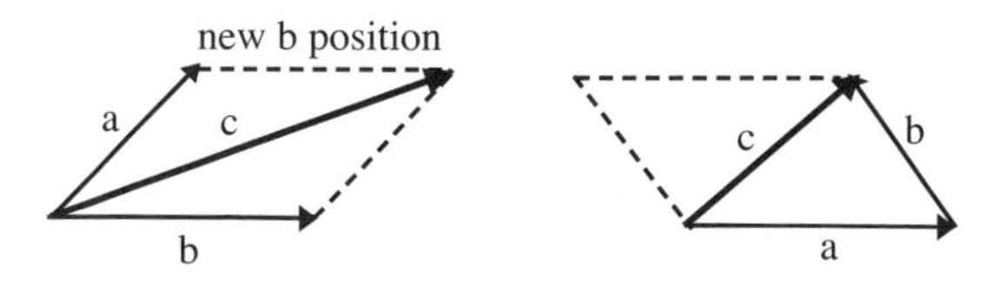

Both examples represent **a** + **b** = **c**

Note that the sum is also the diagonal of a parallelogram that can be constructed on **a** and **b**.

- Subtraction of two vectors is equivalent to addition of the first with the negative of the second vector.

- The negative of a vector is a vector with the same length but pointing in the opposite direction.

- To subtract two vectors, reverse the direction of the second vector, then add the first vector with the negative of the second vector by positioning the vectors so that the initial point of the (negative) second vector is at the final point of the first vector. The sum of the two vectors will be a third vector with its initial point at the initial point of the first vector and its final point at the final point of the second (negative) vector.

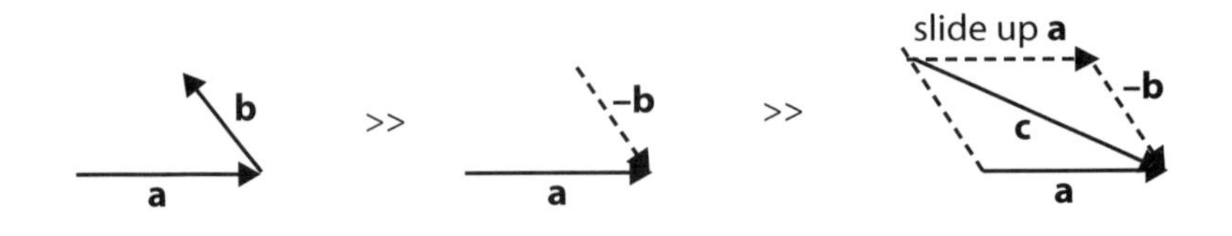

This example represents $\mathbf{a} - \mathbf{b} = \mathbf{c}$

Chapter 2

Trigonometry

2.1 Introduction

• Trigonometry involves the measurement of triangles. Trigonometry includes the measurement of angles, lengths, and arc lengths of triangles in circles and planes and also in spheres. Trigonometry is used in engineering, navigation, the study of electricity, light and sound, and in any field involving the study of periodic and wave properties.

• *Trigonometric functions* can be defined using ratios of sides of a right triangle and, more generally, using the coordinates of points on a circle of radius one. Trigonometric functions are sometimes called *circular functions* because their domains are lengths of arcs on a circle. *Sine, cosine, tangent, cotangent, secant,* and *cosecant* are trigonometric functions. For example, for a circle having a radius of one:

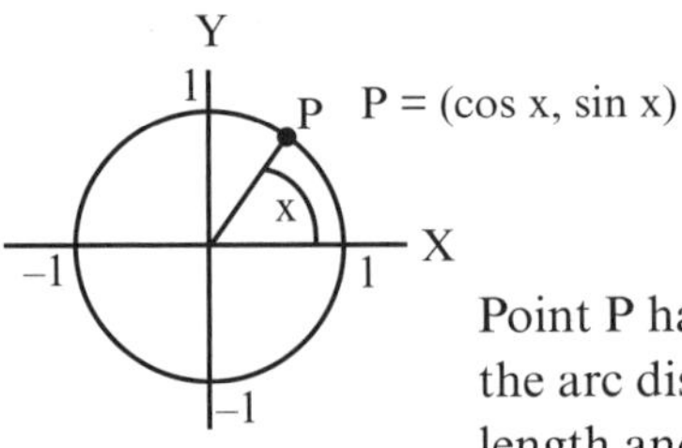

Point P has coordinates (cos x, sin x), and the arc distance of the angle has x units of length and is measured in radians.

• In this chapter trigonometric functions, identities, definitions, graphs, and relationships are presented.

2.2 General Trigonometric Functions

• Certain trigonometric functions describe triangles formed in coordinate systems. For a rectangular coordinate system with an X axis and Y axis, an angle with its vertex at x = 0, y = 0 can be drawn. An angle is said to be in "standard position" if its vertex is at (0, 0) of the X-Y coordinate system and if one side lies on the positive side of the X axis.

• If the standard position angle is measured in a counter-clockwise direction, it is positive. If the standard position angle is measured in a clockwise direction, it is negative.

• The angle, depending on the direction it is measured, has an "initial side" where its measurement begins and a "terminal side" where its measurement ends.

• The following four figures are examples of standard position angles with their terminal sides in the upper right, upper left, lower right, and lower left quadrants.

Standard Position Angle, Ø

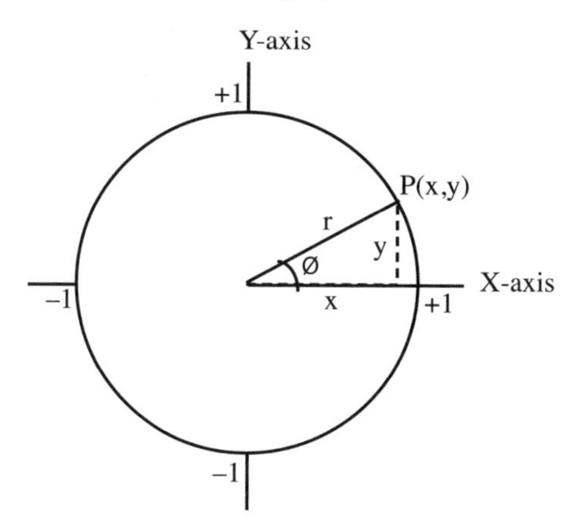

Standard Position Angle, Ø

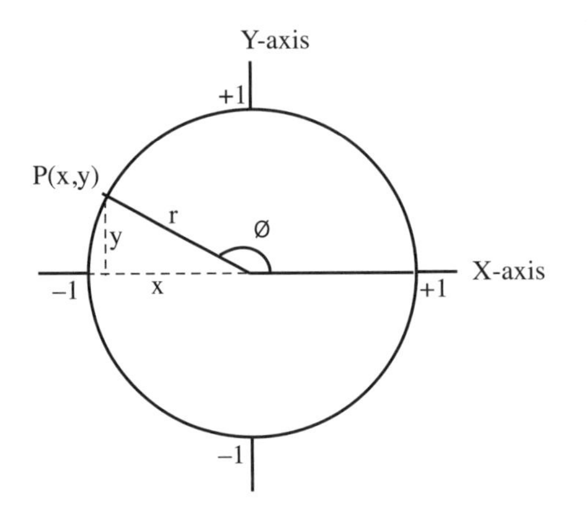

Standard Position Angle, Ø

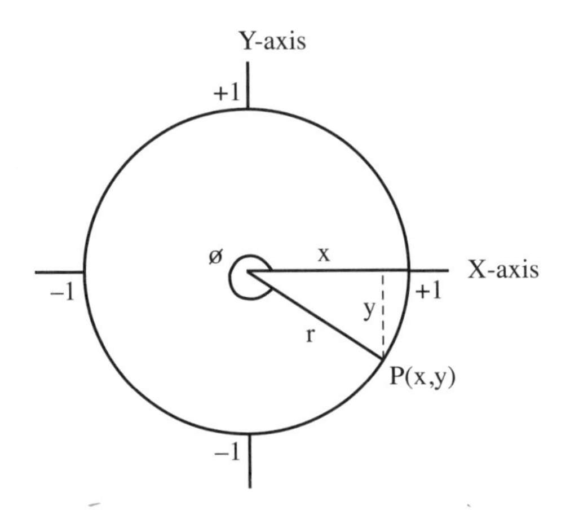

Standard Position Angle, Ø

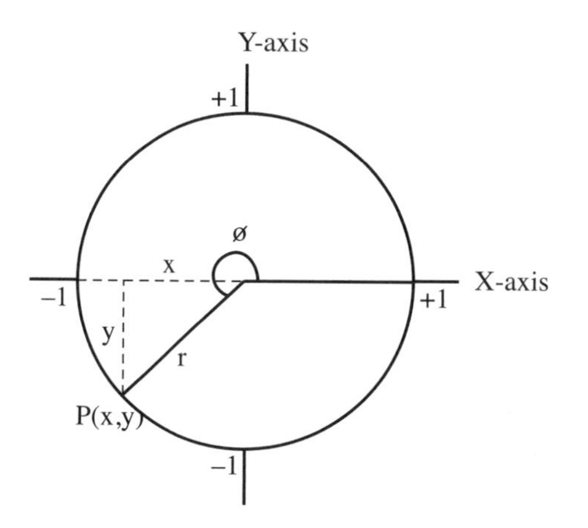

• *Trigonometric functions* defined in terms of the angle (Ø), x, y, and radius r are:

sine Ø = sin Ø = y/r

cosine Ø = cos Ø = x/r

tangent Ø = tan Ø = y/x

cosecant Ø = csc Ø = r/y

secant Ø = sec Ø = r/x

cotangent Ø = cot Ø = x/y

• *Trigonometric identities* are:

cos Ø = 1/ sec Ø

tan Ø = 1/ cot Ø

sec Ø = 1/ cos Ø

cot Ø = 1/ tan Ø

cos Ø tan Ø = sin Ø

$1 + \tan^2 Ø = \sec^2 Ø$

$\sin^2 Ø + \cos^2 Ø = 1$

$\sin Ø = 1/ \csc Ø$

$\tan Ø = \sin Ø/ \cos Ø$

$\csc Ø = 1/ \sin Ø$

$\cot Ø = \cos Ø / \sin Ø$

$\cos Ø \csc Ø = \cot Ø$

$1 + \cot^2 Ø = \csc^2 Ø$

- *Trigonometric functions* can be defined by considering the right triangle formed. If r is the hypotenuse, y is the side opposite Ø, and x is the side adjacent to Ø, then the following is true:

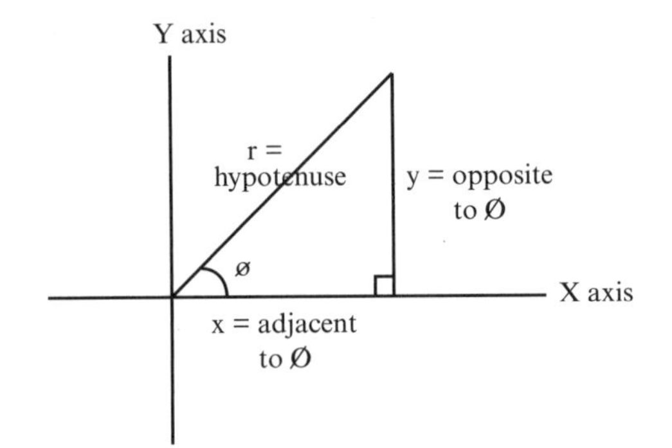

$\sin Ø$ = opposite/hypotenuse = y/r

$\csc Ø$ = hypotenuse/opposite = r/y

$\cos Ø$ = adjacent/hypotenuse = x/r

$\sec Ø$ = hypotenuse/adjacent = r/x

$\tan Ø$ = opposite/adjacent = y/x

$\cot Ø$ = adjacent/opposite = x/y

$(y/r)^2 + (x/r)^2 = 1$ or $y^2 + x^2 = r^2$

Note that the Pythagorean Theorem is $r^2 = x^2 + y^2$
(To remember sin Ø, cos Ø, and tan Ø, think of the word SohCahToa or $S^o{}_h C^a{}_h T^o{}_a$ or $S^o/_h C^a/_h T^o/_a$.)

- Examples of trigonometric functions, ($\pi \approx 3.14$):

degrees	radians	sin	cos	tan	csc	sec	cot
0	0	0	1	0	*	1	*
30	$\pi/6$	1/2	$\sqrt{3}/2$	$\sqrt{3}/3$	2	$2\sqrt{3}/3$	$\sqrt{3}$
45	$\pi/4$	$\sqrt{2}/2$	$\sqrt{2}/2$	1	$\sqrt{2}$	$\sqrt{2}$	1
60	$\pi/3$	$\sqrt{3}/2$	1/2	$\sqrt{3}$	$2\sqrt{3}/3$	2	$\sqrt{3}/3$
90	$\pi/2$	1	0	*	1	*	0
180	π	0	–1	0	*	–1	*
270	$3\pi/2$	–1	0	*	–1	*	0
360	2π	0	1	0	*	1	*

* = undefined

(See a trigonometry book, calculus book, or book of mathematical tables for more values.)

- *Trigonometric functions* of angles of a 30:60:90 triangle. (a = adjacent, o = opposite, h = hypotenuse)

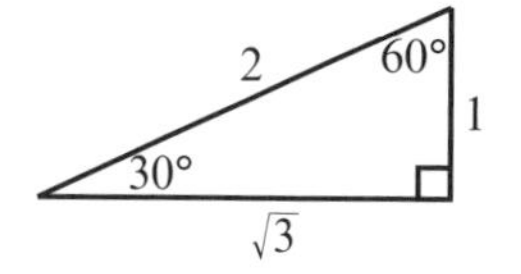

30 degrees = $\pi/6$ radians
60 degrees = $\pi/3$ radians
90 degrees = $\pi/2$ radians
$\cos 30° = \cos(\pi/6) = a/h = \sqrt{3}/2$
$\sin 30° = \sin(\pi/6) = o/h = 1/2$
$\tan 30° = \tan(\pi/6) = o/a = 1/\sqrt{3} = \sqrt{3}/3$
$\cos 60° = \cos(\pi/3) = a/h = 1/2$
$\sin 60° = \sin(\pi/3) = o/h = \sqrt{3}/2$
$\tan 60° = \tan(\pi/3) = o/a = \sqrt{3}$

Remember 2π radians = 360°. This can be used as a conversion factor when transforming degrees to radians.

• Some basic properties and formulas of the trigonometric functions are:

Sine is an odd function; therefore, for any number Ø
$\sin(-Ø) = -\sin Ø$

Cosine is an even function; therefore, for any number Ø
$\cos(-Ø) = \cos Ø$

Tangent is an odd function; therefore, for any number Ø
$\tan(-Ø) = -\tan Ø$

$\cot(-Ø) = -\cot Ø$
$\sec(-Ø) = \sec Ø$
$\csc(-Ø) = -\csc Ø$

$\sin(\pi/2 + Ø) = \cos Ø$
$\sin(\pi/2 - Ø) = \cos Ø$

$\cos(\pi/2 + Ø) = -\sin Ø$
$\cos(\pi/2 - Ø) = \sin Ø$

$\sin(90° + Ø) = \cos Ø$
$\sin(90° - Ø) = \cos Ø$

$\cos(90° + Ø) = -\sin Ø$
$\cos(90° - Ø) = \sin Ø$

$\tan(90° + Ø) = -\cot Ø$
$\tan(90° - Ø) = \cot Ø$

$\sin(180° + Ø) = -\sin Ø$
$\sin(180° - Ø) = \sin Ø$

$\cos(180° + Ø) = -\cos Ø$
$\cos(180° - Ø) = -\cos Ø$

$\tan (180° + Ø) = \tan Ø$
$\tan (180° - Ø) = -\tan Ø$

$\sec (90° + Ø) = -\csc Ø$
$\sec (90° - Ø) = \csc Ø$

$\csc (90° + Ø) = \sec Ø$
$\csc (90° - Ø) = \sec Ø$

$\cot (90° + Ø) = -\tan Ø$
$\cot (90° - Ø) = \tan Ø$

$\sec (180° + Ø) = -\sec Ø$
$\sec (180° - Ø) = -\sec Ø$

$\csc (180° + Ø) = -\csc Ø$
$\csc (180° - Ø) = \csc Ø$

$\cot (180° + Ø) = \cot Ø$
$\cot (180° - Ø) = -\cot Ø$

$\sin 2 Ø = 2 \sin Ø \cos Ø$
$\cos 2 Ø = 2 \cos^2 Ø - 1$
$\tan 2 Ø = 2 \tan Ø / 1 - \tan^2 Ø$

Sum, difference, and product formulas:

$\cos x - \cos y = -2 \sin(1/2)(x + y) \sin(1/2)(x - y)$
$\cos x + \cos y = 2 \cos(1/2)(x + y) \cos(1/2)(x - y)$
$\sin x - \sin y = 2 \cos(1/2)(x + y) \sin(1/2)(x - y)$
$\sin x + \sin y = 2 \sin(1/2)(x + y) \cos(1/2)(x - y)$
$\sin x \cos y = (1/2)\sin(x - y) + (1/2)\sin(x + y)$
$\cos x \sin y = (1/2)\sin(x + y) - (1/2)\sin(x - y)$
$\cos x \cos y = (1/2)\cos(x - y) + (1/2)\cos(x + y)$
$\sin x \sin y = (1/2)\cos(x - y) - (1/2)\cos(x + y)$

2.3 Addition, Subtraction, and Multiplication of Two Angles

- The following identities for two adjacent angles (Ø and Ω) are:

Angles Ø and Ω

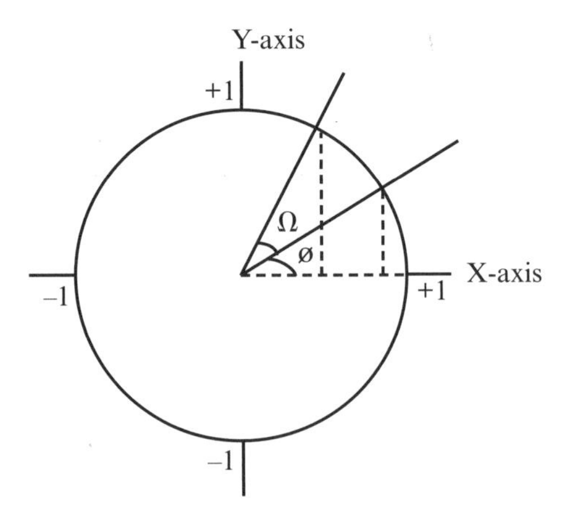

$\sin(Ø + \Omega) = \sin Ø \cos \Omega + \cos Ø \sin \Omega$

$\cos(Ø + \Omega) = \cos Ø \cos \Omega - \sin Ø \sin \Omega$

$\tan(Ø + \Omega) = (\tan Ø + \tan \Omega) / (1 - \tan Ø \tan \Omega)$

$\sin(Ø - \Omega) = \sin Ø \cos \Omega - \cos Ø \sin \Omega$

$\cos(Ø - \Omega) = \cos Ø \cos \Omega + \sin Ø \sin \Omega$

$\tan(Ø - \Omega) = (\tan Ø - \tan \Omega) / (1 + \tan Ø \tan \Omega)$

$\sin Ø \cos \Omega = (1/2)(\sin(Ø + \Omega) + \sin(Ø - \Omega))$

$\cos Ø \sin \Omega = (1/2)(\sin(Ø + \Omega) - \sin(Ø - \Omega))$

$\cos Ø \cos \Omega = (1/2)(\cos(Ø + \Omega) + \cos(Ø - \Omega))$

$\sin Ø \sin \Omega = (-1/2)(\cos(Ø + \Omega) - \cos(Ø - \Omega))$

2.4 Oblique Triangles

• *Oblique triangles* are triangles in planes that are not right triangles. They are described using the *Law of Sines, Law of Cosines*, and *Law of Tangents.*

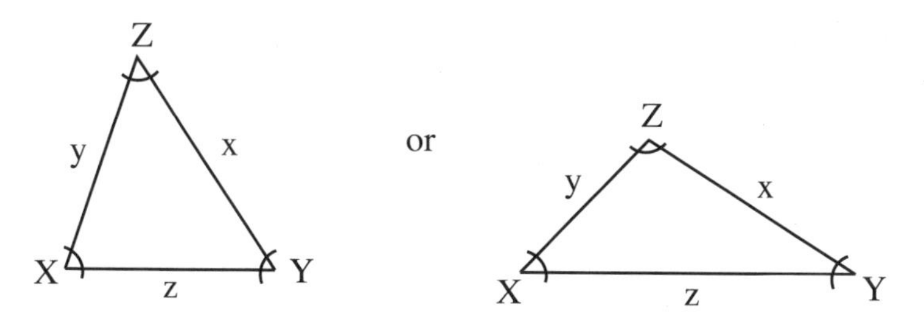

Law of Sines

$$\frac{x}{\sin X} = \frac{y}{\sin Y} = \frac{z}{\sin Z}$$

By rearranging, the following are obtained:

$$\frac{x}{y} = \frac{\sin X}{\sin Y}, \quad \frac{y}{z} = \frac{\sin Y}{\sin Z}, \quad \frac{z}{x} = \frac{\sin Z}{\sin X}$$

Law of Cosines

$$x^2 = y^2 + z^2 - 2yz \cos X$$
$$y^2 = z^2 + x^2 - 2zx \cos Y$$
$$z^2 = x^2 + y^2 - 2xy \cos Z$$

Law of Tangents

$$\frac{x + y}{x - y} = \frac{\tan \frac{1}{2}(X + Y)}{\tan \frac{1}{2}(X - Y)}$$

2.5 Graphs of Cosine, Sine, Tangent, Secant, Cosecant, and Cotangent

• The graphs in this section depict cosine, sine, tangent, secant, cosecant, and cotangent. Cosine, sine, and tangent are described by the following equations:

$\cos x = \cos(x + 2n\pi)$

$\cos x = \sin(\pi/2 + x)$

$\sin x = \sin(x + 2n\pi)$

$\tan x = \tan(x + n\pi)$

Where n is any integer and x is any real number.

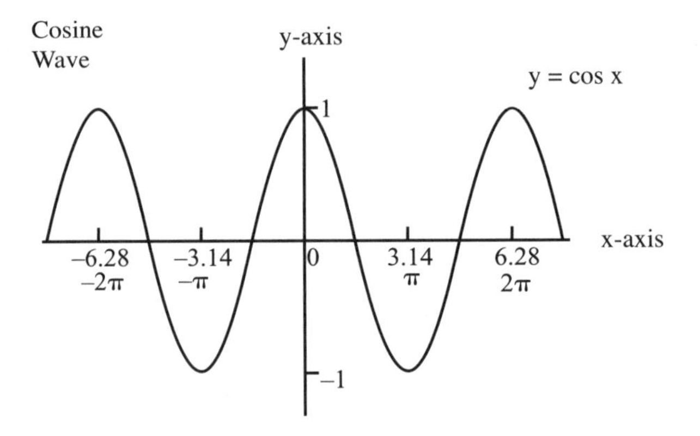

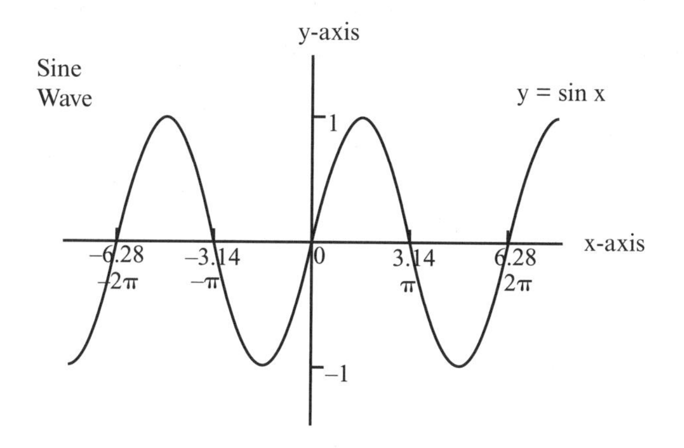

It is possible to graph y = cos x and y = sin x by selecting values for x and calculating the corresponding y values.

If there are coefficients in the equations y = cos x and y = sin x, the function will have the same general shape, but it will have a larger or smaller amplitude (taller or shorter), or it will be elongated or narrower, or it will be moved to the right or left or up and down.

For example, if there is a coefficient of 2 in front of cos or sin, the graph will go to +2 and −2 (rather than +1 and −1) on the y axis.

Similarly, if there is a coefficient of 1/2 in front of cos or sin, the graph will go to +1/2 and −1/2 (rather than +1 and −1) on the y axis.

If there is a coefficient of 2 in front of x, giving y = cos 2x and y = sin 2x, the graph will complete each cycle along the x axis twice as fast. Because there is one cycle between 0 and 2π for y = cos x and y = sin x, there will be two cycles between 0 and 2π for y = cos 2x and y = sin 2x.

Similarly, if there is a coefficient of 1/2 in front of x, giving y = cos x/2 and y = sin x/2, the graph will complete each cycle along the x axis half as fast. Because there is one cycle between 0 and 2π for y = cos x and y = sin x, there will be one-half of a cycle between 0 and 2π for y = cos 2x and y = sin 2x.

If a number is added or subtracted giving, for example, y = cos x + 2 and y = sin x + 2, the function will be moved up or down on the y axis, in this case, up 2.

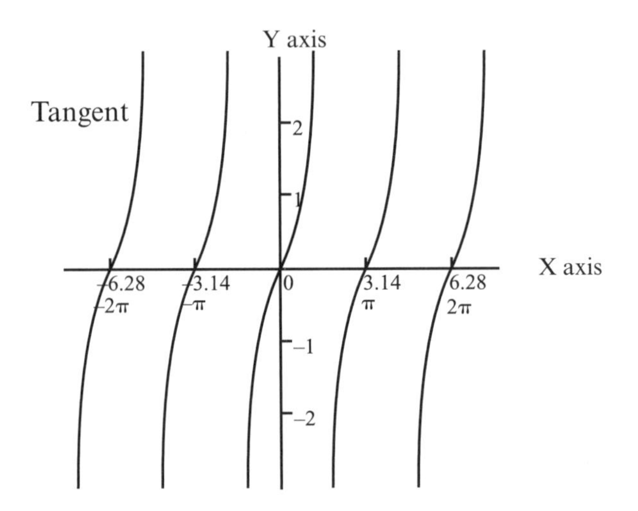

• The following graphs depict secant, cosecant, and cotangent:

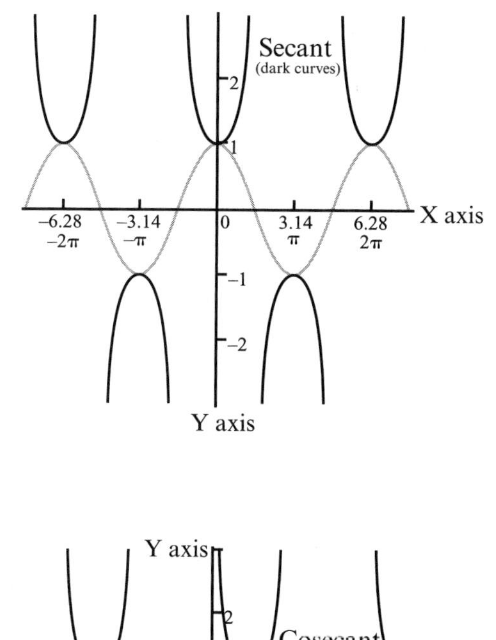

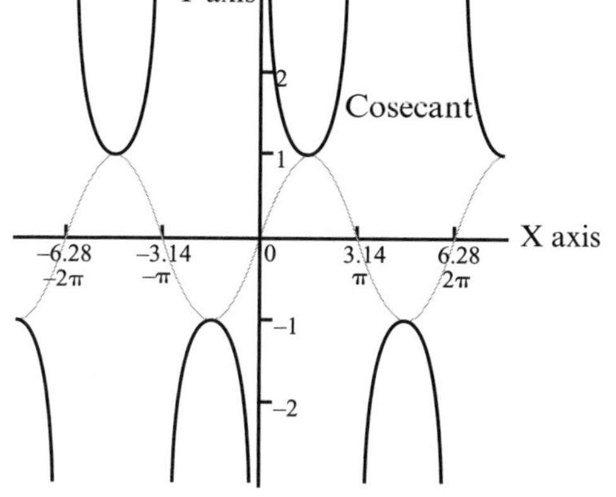

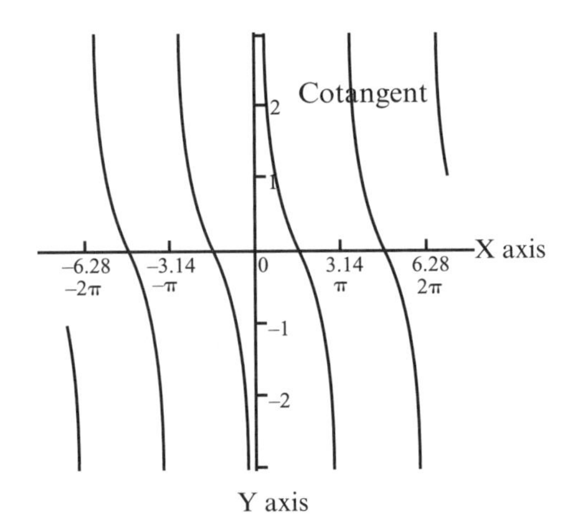

2.6 Relationship Between Trigonometric and Exponential Functions

• Trigonometric functions and exponential functions are related to each other. The following equations define the relationship between these functions.

• Note that $i = \sqrt{-1}$. See imaginary numbers in Section 1.17, "Complex Numbers," in *Basic Math and Pre-Algebra,* the first book in the *Master Math* series.

$$e^{i\theta} = \cos\theta + i\sin\theta$$

(This is Euler's identity. It defines the simple relationship between $e^{i\theta}$, $\cos\theta$, and $\sin\theta$.)

$$e^{-i\theta} = \cos\theta - i\sin\theta$$

$$e^{i(-\theta)} = \cos(-\theta) + i\sin(-\theta)$$

$$\cos\theta = \frac{e^{i\theta} + e^{-i\theta}}{2}$$

$$\sin\theta = \frac{e^{i\theta} - e^{-i\theta}}{2i}$$

The expansions for e^x, cos x, and sin x are:

$$e^x = 1 + x + \frac{x^2}{2!} + \frac{x^3}{3!} + \frac{x^4}{4!} + \cdots + \frac{x^n}{n!} + \cdots$$

$$\cos x = 1 - \frac{x^2}{2!} + \frac{x^4}{4!} - \cdots$$

$$\sin x = x - \frac{x^3}{3!} + \frac{x^5}{5!} - \cdots$$

Therefore, the expansions for $e^{i\theta}$, cos θ, and sin θ are:

$$e^{i\theta} = 1 + i\theta - \frac{\theta^2}{2!} - i\frac{\theta^3}{3!} + \frac{\theta^4}{4!} + \cdots + \frac{\theta^n}{n!} + \cdots$$

$$\cos\theta = 1 - \frac{\theta^2}{2!} + \frac{\theta^4}{4!} - \cdots$$

$$i\sin\theta = i\theta - \frac{i\theta^3}{3!} + \frac{i\theta^5}{5!} - \cdots$$

2.7 Hyperbolic Functions

- Hyperbolic functions are real and are derived from the exponential functions e^{θ} and $e^{-\theta}$. The following equations define hyperbolic functions.

The hyperbolic cosine:

$$\cosh\theta = \frac{e^{\theta} + e^{-\theta}}{2}$$

The hyperbolic sine:

$$\sinh\theta = \frac{e^{\theta} - e^{-\theta}}{2}$$

The hyperbolic tangent:

$$\tanh\theta = \frac{\sinh\theta}{\cosh\theta} = \frac{e^{\theta} - e^{-\theta}}{e^{\theta} + e^{-\theta}}$$

The hyperbolic cosecant:

$$\operatorname{csch}\theta = \frac{1}{\sinh\theta}$$

The hyperbolic secant:

$$\operatorname{sech}\theta = \frac{1}{\cosh\theta}$$

The hyperbolic cotangent:

$$\coth\theta = \frac{1}{\tanh\theta}$$

- Note that $\cosh\theta$ and $\sinh\theta$ are similar to the functions for $\cos\theta$ and $\sin\theta$.

$$\cos\theta = \frac{e^{i\theta} + e^{-i\theta}}{2} \quad \text{and} \quad \sin\theta = \frac{e^{i\theta} - e^{-i\theta}}{2i}$$

Chapter 3

Sets and Functions

3.1 Sets

- Any group or collection of objects can form a *set*. The objects in a set are called *elements.*

- For example, in the set of numbers:

$$S = \{3, 5, 10, 17\}$$

S depicts the set, and 3, 5, 10, and 17 are its elements.

- Because 5 is a member of the Set S, we write:

$$5 \in S$$

The symbol for member is "$\in$".

- Q is a set where, $Q = \{5, 10\}$.

Note that Q is also a subset of Set S.

- The symbol for *subset* is "$\subset$", where $Q \subset S$.
- If P is an *empty set,* we write $P = \{\}$.

An empty set is also called a *null set.*

- An empty set is a subset of every other set, so the following is true:

$$P \subset Q \text{ and } P \subset S$$

- An alternative method for describing a set is to write:

$$\{x: \qquad \}$$

This set contains values for x, which are described by what is written after the colon.

- For example:

$$\{x: x \text{ is a quadrilateral}\}$$

All members of this set are quadrilaterals (planar four-sided objects), including squares, rectangles, parallelograms, and trapezoids.

- The arrangement of the objects or values in the set does not matter.

$$Q = \{5, 10\} = \{10, 5\}$$

• Duplicates or replicates of an object or value in a set do not count as extra objects or values and do not change the net value of the set.

$Q = \{5, 10\} = \{5, 10, 5\} = \{5, 10, 10\} = \{10, 5, 5, 5, 10\}$

• Combining sets is called a *union* of the sets. The symbol for union is "∪".

$S \cup Q = \{3, 5, 10, 17\} \cup \{5, 10\}$

$= \{3, 5, 10, 17, 5, 10\} = \{3, 5, 10, 17\}$

• Suppose there are two sets and some of the objects or values in one set are the same as some of the objects or values in the other set. If the objects that are common to both sets are placed in a new set, the new set is called an *intersection* of the first two sets. The symbol for intersection is "∩".

Set $X = \{\pi, 2, 7\}$

Set $Y = \{2, 5, 8\}$

$X \cap Y = \{2\}$

Venn diagrams are often used to represent sets and their intersection. This is a Venn diagram of $X \cap Y$, where set X contains π, 7, and 2, and set Y contains 5, 8, and 2. The intersection of set X and Y is a third set that contains 2.

$X \cap Y =$

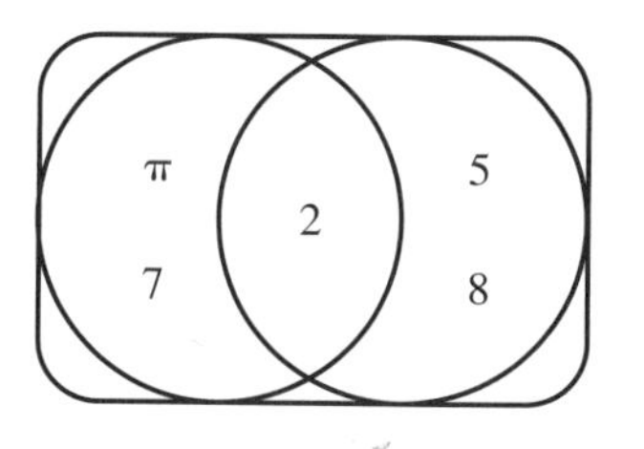

3.2 Functions

• In this section, functions, domain set, range set, graphing functions, compound functions, inverse functions, as well as adding, subtracting, multiplying, and dividing functions are described.

• A *function* is a relation, rule, expression, or equation that associates each element of a *domain set* with its corresponding element in the *range set*. For a relation, rule, expression, or equation to be a function, there must be only one element or number in the range set for each element or number in the domain set.

• The domain set is the initial set and the range set is the resulting set when a function is applied.

Domain Set → Function f() → Range Set

• The function f( ) is like a transformer that translates the domain set into the range set.

• Each time a given member of the domain set goes through the function transformer, it always produces the same member in the range set.

domain set x → function (transformer) f( )
→ range set f(x)

For example, domain set $x = \{2, 3, 4\}$

through function $f(x) = x^2$,

Where $f(2) = 2^2$, $f(3) = 3^2$, and $f(4) = 4^2$

to range set $f(x) = \{4, 9, 16\}$

• The domain set and range set can be expressed as $(x, f(x))$ pairs. In the previous example, the function is $f(x) = x^2$ and the pairs are (2, 4), (3, 9), and (4, 16).

• For each member of the domain set, there must be only one corresponding member in the range set. For example:

$F = (2, 4), (3, 9), (4, 16)$ Where F is a function.

$M = (2,5), (2, -5), (4, 9)$ Where M is not a function.

M is not a function because the number 2 in the domain set corresponds to more than one number in the range set.

• To *graph functions*, the values in the domain set correspond to the X-axis and the related values in the range set correspond to the Y-axis.

domain set $x = -2, -1, 0, 2$

through function $f(x) = x + 1$

to range set $f(x) = -1, 0, 1, 3$

resulting in pairs $(x, y) = (-2, -1), (-1, 0), (0, 1), (2, 3)$

When graphed these resulting pairs are depicted as:

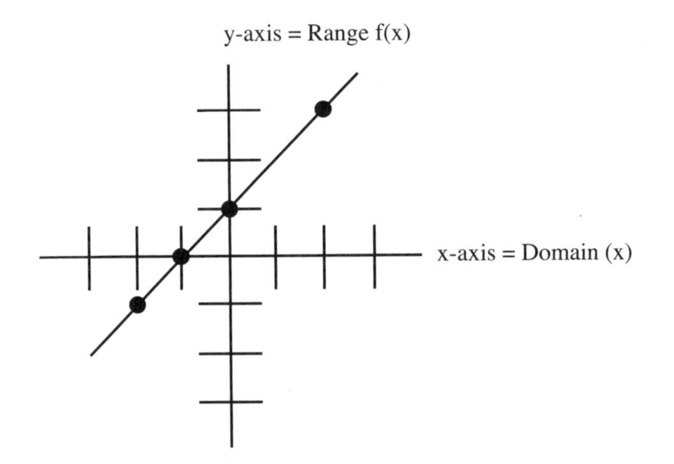

• Graphs of functions only have one value for y for each x value.

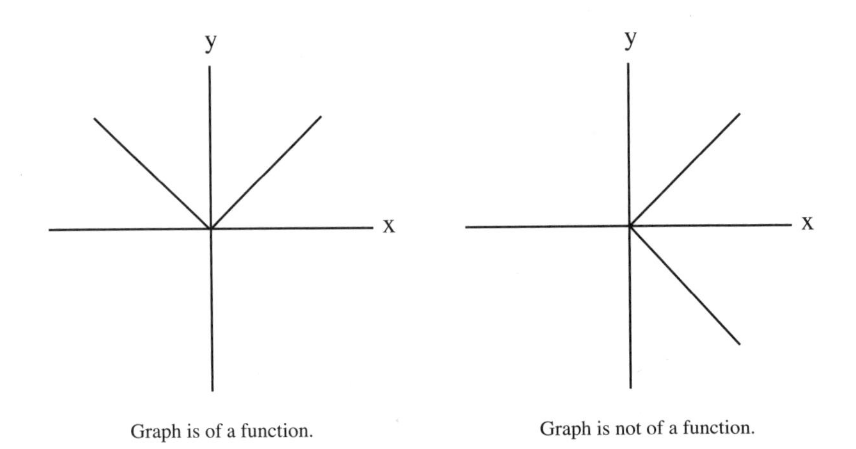

Graph is of a function. Graph is not of a function.

If a vertical line can be drawn that passes through the function more than one time, there is more than one y value for a given x value and the graph is not a function. This is called the *vertical line test.*

• The following are examples of addition, subtraction, multiplication, and division of functions. In these examples the functions f(x) and g(x) are given by:

$f(x) = 2x$ and $g(x) = x^2$

Add two functions:

$$f(x) + g(x) = (f + g)(x) = 2x + x^2$$

Subtract two functions:

$$f(x) - g(x) = (f - g)(x) = 2x - x^2$$

Multiply two functions:

$$f(x) \times g(x) = (f \times g)(x) = 2x \times x^2 = 2x^3$$

Divide two functions:

$$f(x) \div g(x) = (f \div g)(x) = 2x \div x^2 = \frac{2x}{x^2} = \frac{2}{x}$$

• *Compound functions* are functions that are combined, and the operations specified by the functions are combined. Compound functions are written:

$f(g(x))$ or $g(f(x))$

If $f(x) = x + 1$ and $g(x) = 2x - 2$,
then the compound functions for f(g(x)) and g(f(x)) are:

$$f(g(x)) = f(2x - 2) = (2x - 2) + 1 = 2x - 1 \text{ and}$$

$$g(f(x)) = g(x + 1) = 2(x + 1) - 2 = 2x + 2 - 2 = 2x$$

The notation for compound functions may also be written:

$f \circ g$ and $g \circ f$

• *Inverse functions* are functions that result in the same value of x after the operations of the two functions are performed. In inverse functions the operations of each function are the reverse of the other function. The notation for inverse is $f^{-1}(x)$. Functions are inverse if:

$$f(g(x)) = x \text{ and } g(f(x)) = x$$

• For example, if $f(x) = 2x - 1$, find the inverse by solving for x.

Let $f(x) = y$, the function becomes:

$$y = 2x - 1$$

$$y + 1 = 2x$$

$$(y + 1)/2 = x$$

If y and x are switched, the inverse of $f(x) = 2x - 1$ is

$$f^{-1}(x) = (x + 1)/2$$

To verify that $f(x)$ and $f^{-1}(x)$ are inverse functions, let $x = 3$ and substitute into the functions.

$$f(x) = 2x - 1 = f(3) = 2(3) - 1 = 6 - 1 = 5$$

Substitute the result, 5, into $f^{-1}(x)$

$$f^{-1}(x) = (x + 1)/2 = f^{-1}(5) = (5 + 1)/2 = 6/2 = 3$$

The result is 3.

Also, verify the inverse using $f(f^{-1}(x)) = 2[(x + 1)/2] - 1$:

$$\text{If } x = 3,\ f(f^{-1}(3)) = 2[(3 + 1)/2] - 1 = 2[2] - 1 = 3$$

Or using $f^{-1}(f(x)) = ([2x - 1] + 1)/2$:

$$\text{If } x = 3,\ f^{-1}(f(3)) = ([2(3) - 1] + 1)/2 = ([5] + 1)/2 = 3$$

Chapter 4

Sequences, Progressions, and Series

4.1 Sequences

• A *sequence* is a set of numbers called terms, which are arranged in a succession in which there is a relationship or rule between each successive number.

• A sequence can be finite or infinite. A *finite sequence* has a last term, and an *infinite sequence* has no last term.

• The following is an example of a finite sequence:

{3, 6, 9, 12, 15, 18}

In this sequence, each number has a value of 3 more than the preceding number.

• The following is an example of an infinite sequence describing the function f(x) = 1/x:

{1/1, 1/2, 1/3, 1/4, 1/5, 1/6,...}
domain set is x = {1, 2, 3, 4, 5, 6,...} and the
range set is f(x) = {1/1, 1/2, 1/3, 1/4, 1/5, 1/6,...}

• For example, the *Fibonacci sequence* is an unending sequence where each term is defined as the sum of its two predecessors. The numbers in the Fibonacci sequence are called the Fibonacci numbers and have many applications, including the pattern of the spiral curve arrangement of the seed head of a sunflower. Following is the Fibonacci sequence:

{1, 1, 2, 3, 5, 8, 13, 21, 34, ...}

4.2 Arithmetic Progressions

• An *arithmetic progression* is a sequence in which the difference between successive terms is a fixed number, and each term is obtained by adding a fixed amount to the term before it. This fixed amount is called the *common difference*.

• Arithmetic progressions can be represented by first-degree polynomial expressions. For example, the expression $n+1$ can represent an arithmetic progression.

• The first sequence example in the previous section is an arithmetic progression:

$$\{3, 6, 9, 12, 15, 18\}$$

This arithmetic progression is represented by:

$$n + 3$$

• A finite arithmetic progression can be expressed as:

$$a, a + d, a + 2d, a + 3d, a + 4d, a + 5d, \ldots, a + (n - 1)d$$

a is the first term.

d is the fixed difference between each term.

$a + (n - 1)d$ is the last term or "nth" term.

Each term can be written as:

$n = 1, a_1 = a + (1 - 1)d = a$

$n = 2, a_2 = a + (2 - 1)d = a + d$

$n = 3, a_3 = a + (3 - 1)d = a + 2d$

$n = 4, a_4 = a + (4 - 1)d = a + 3d$

$n = 5, a_5 = a + (5 - 1)d = a + 4d$

and so on.

- In the arithmetic progression on the previous page:

$\{3, 6, 9, 12, 15, 18\}$,
$a = 3$ and $d = 3$.

Therefore,

for $n = 1, a_1 = 3$,
for $n = 2, a_2 = 6$,
for $n = 3, a_3 = 9$,
and so on.

4.3 Geometric Progressions

- A *geometric progression* is a sequence in which the ratio of successive terms is a fixed number, and each term is obtained by multiplying a fixed amount to the term before it. This fixed amount is called the *common ratio*.

- The terms in a geometric progression can be represented as:

$a, ar, ar^2, ar^3, ar^4, ar^5, \ldots, ar^{n-1}$

a is the first term.
r is the ratio between successive terms.
ar^{n-1} is the last term.

- Note that the ratio of successive terms is always r.

$ar/a = r$
$ar^2/ar = r$
$ar^3/ar^2 = r$

- For example, consider the following geometric progression:

2, 4, 8, 16, 32,...

If $a = 2$ and $r = 2$

This geometric progression can be expressed as:

$2, 2(2), 2(2)^2, 2(2)^3, 2(2)^4, \ldots, 2(2)^{n-1}$

4.4 Series

- A *series* is the sum of the terms in a progression or sequence. An *arithmetic series* is the sum of the terms in an arithmetic progression. A *geometric series* is the sum of the terms in a geometric progression. The notation used to express a series is *sigma notation.*

- The sigma notation that represents an arithmetic series is:

$$\sum_{n=1}^{m} a_n$$

a_n is the sequence function.

m is the index of the last term that is added.

n represents the nth term.

- For example, to obtain the sum of the first three terms in the sequence a_n from n = 1 to n = 3, the sigma notation is:

$$\sum_{n=1}^{3} a_n = a_1 + a_2 + a_3$$

- In the arithmetic progression {3, 6, 9, 12,...}, for the first three terms:

$$\sum_{n=1}^{3} a_n = 3 + 6 + 9 = 18$$

- To sum a large number of terms in an arithmetic progression, the following equation for the arithmetic series can be used:

$$(m/2)(a_1 + a_m)$$

Where m represents the last term added.

- Apply the equation for the arithmetic series to the arithmetic progression {3, 6, 9}.

$$(3/2)(3 + 9) = (3/2)(12) = 18$$

- Sigma notation that represents a geometric series is:

$$\sum_{n=1}^{m} ar^{n-1} = a + ar + ar^{2} + ar^{3} + ar^{4} + ar^{5} + \cdots + ar^{n-1}$$

a is the first term and $a \neq 0$.

r is the ratio between successive terms.

m is the index of the last term added.

n represents the nth term.

ar^{n-1} is the last term.

- In the geometric progression {2, 4, 8, 16, 32,...}, the sum of the first three terms is:

$$2 + 4 + 8 = 14$$

- The sum of a large number of terms in a geometric progression can be represented by:

$$\frac{a(1-r^{m})}{(1-r)}$$

Where m represents the last term added and r is the ratio.

- Apply the equation for the geometric series to the geometric progression {2, 4, 8}.

$$\frac{2(1-2^{3})}{1-2} = \frac{2(1-8)}{1-2} = \frac{2(-7)}{-1} = \frac{-14}{-1} = 14$$

• If the geometric series is infinite, m will approach infinity (∞), therefore, the limit as m approaches infinity of the expression of the series is:

$$\text{Lim}_{m->\infty} \frac{a(1-r^m)}{(1-r)}$$

If $|r| < 1$ and $m \to \infty$, then r^m approaches zero, and the sum of the infinite geometric series becomes:

$$\frac{a}{(1-r)}$$

(See the next section for proof.)

4.5 Infinite Series: Convergence and Divergence

• A *series* is a sum of the terms in a progression or sequence. If the progression or sequence is infinite and therefore there is an infinite number of terms, then the sum cannot be calculated. However, under certain conditions the sum of an *infinite series* can be estimated.

• Conditions under which the sum of an infinite series can be estimated include the following.

If an infinite series has a limit, it is said to *converge*, and the sum can be estimated. In other words, as the terms in an infinite series are added, beginning with the first term, if with each additional term added the sum approaches some number, then the series has a limit and converges, and the sum can be estimated.

If an infinite series has no limit, it is said to *diverge*, and the sum cannot be estimated. In other words, if instead, as each additional term is added the sum approaches infinity, then the series has no limit and diverges, and the sum cannot be estimated.

Another condition for convergence is that, for the infinite series that follows:

$$\sum_{n=1}^{\infty} a_n$$

a_n must approach zero as n approaches infinity.

Although this condition must occur for a series to converge, there are cases where this condition is true but the series still diverges.

- To estimate an infinite series, it must be determined whether the series has a limit and converges, and what happens to the sum as the number of terms approach infinity. For example, consider the infinite series describing the sum of a_n from $n = 1$ to $n = \infty$:

$$\sum_{n=1}^{\infty} a_n$$

If this series has a limit and converges to L it can be written:

$$\text{Lim}_{n\to\infty} \sum_{n=1}^{\infty} a_n = L$$

• The geometric series $a + ar + ar^2 + ar^3 + ar^4 + ar^5 + \ldots + ar^{n-1}$, converges when $|r| < 1$ and diverges when $|r| \geq 1$. The geometric series can be expressed as:

$$\sum_{n=1}^{m} ar^{n-1} = a + ar + ar^2 + ar^3 + ar^4 + \cdots + ar^{n-1}$$

and the sum of the first m terms is:

$$\frac{a(1-r^m)}{1-r} = \frac{a - ar^m}{1-r} = \frac{a}{1-r} - \frac{ar^m}{1-r}$$

As m approaches infinity:

$$\mathrm{Lim}_{m\to\infty}\left(\frac{a}{1-r} - \frac{ar^m}{1-r}\right)$$

If $|r| < 1$ and $m\to\infty$, then $r^m \to 0$, resulting in:

$$\mathrm{Lim}_{m\to\infty}\left(\frac{a}{1-r} - \frac{ar^m}{1-r}\right) = \frac{a}{1-r}$$

Therefore, as $m\to\infty$, if $|r| < 1$, the series converges, and if $|r| \geq 1$, it diverges.

• To determine whether a given series will converge, there are a variety of tests for convergence that may be used, including the Comparison Test, the Ratio Test, and the Integral Test. These three tests are discussed in the following section.

• Note: The *limit* is used in the estimation of infinite sums and the definition of the derivative in calculus. In general, the limit is used to describe *closeness* of a function to a value. For example, if the $\text{Lim}_{x \to a} f(x) = L$, then it is said that the function f(x) gets close to and may equal some number L as x approaches and gets close to some number "a". (Please refer to Chapter 5, "Limits.")

4.6 Tests for Convergence of Infinite Series

• To determine whether an infinite series will converge, there are a variety of tests for convergence that may be used. These tests include the Comparison Test, the Ratio Test, the Integral Test, and tests for series with positive and negative terms.

The Comparison Test for Convergence

• The *Comparison Test* can be applied to a series with positive terms.

A series is convergent if each term is less than or equal to each corresponding term in a series that is known to be convergent.

Conversely, if each term in an unknown infinite series is greater than or equal to each corresponding term in a known divergent infinite series, then the unknown infinite series is also divergent.

- An example of a known convergent series is the P Series:

$$1 + 1/2^P + 1/3^P + \ldots + 1/n^P + \ldots$$

The P Series converges when $P > 1$ and diverges when $P \leq 1$.

- A divergent series that is used as a comparison in the Comparison Test is:

$$1 + 1 + 1 + 1 + \ldots$$

As the number of terms approaches infinity, the sum of the terms approaches infinity and the series diverges.

- Example: Will Series U converge?

$$U = 1 + 1/2 + 1/3 + \ldots + 1/n + \ldots$$

(This is called the Harmonic Series.)

Compare with the known Series K, which diverges as more terms are added.

$$K = 1 + 1/2 + 1/2 + 1/2 + \ldots$$

To compare Series K to Series U rewrite Series K as:

$$K = 1 + (1/2) + (1/4 + 1/4) + (1/8 + 1/8 + 1/8 + 1/8) + \ldots$$

Compare Series K to Series U term by term.

$$U = 1 + 1/2 + 1/3 + 1/4 + 1/5 + 1/6 + 1/7 + 1/8\ldots$$

Many of the terms in U are greater than the corresponding terms in K. Therefore, because K diverges, U must also diverge.

• This example is an interesting case because in the Harmonic Series, the value of the terms do approach zero, which is a necessary criteria for convergence but does not guarantee it. However, by applying the Comparison Test with a series that is known to diverge, it is clear that the Harmonic Series also diverges.

The Ratio Test for Convergence

• The *Ratio Test* for convergence can be applied to a series of positive terms.

• To apply the Ratio Test for the series:

$$a_1 + a_2 + a_3 + \ldots a_n + \ldots$$

Find the ratio of successive terms:

$$\frac{a_{n+1}}{a_n} = r_n$$

To determine r, take the Limit as $n \rightarrow \infty$:

$$\text{Lim}_{n-\infty} \left(\frac{a_{n+1}}{a_n} \right) = r$$

If $r < 1$, the series will converge.

If $r > 1$, the series will diverge.

If $r = 1$, the test does not indicate convergence or divergence.

• The Ratio Test can be applied to evaluate convergence of series containing positive and negative terms.

• To apply the Ratio Test to an alternating series, take the Limit as $n \to \infty$ for the ratio of the absolute value of successive terms:

$$\text{Lim}_{n \to \infty} \frac{|a_{n+1}|}{|a_n|} = r$$

If $r < 1$, the series is absolutely convergent.

If $r > 1$, the series is divergent.

If $r = 1$, this test does not indicate convergence or divergence.

This series therefore may be absolutely convergent, conditionally convergent, or divergent.

Tests for Series with Positive and Negative Terms

• A *series with positive and negative terms* converges if the corresponding series of the absolute values of the terms converges.

If Series $|S|$ converges then series S will converge.

$$|S| = |a_1| + |a_2| + |a_3| + |a_4| + |a_5| + \ldots |a_n| + \ldots$$

and:

$$S = a_1 + a_2 + a_3 + a_4 + a_5 + \ldots a_n + \ldots$$

a_n can be positive or negative.

• A series with positive and negative terms may converge and is called *conditionally convergent*, even though its corresponding series of absolute values diverges. For example:

$1 - 1/2 + 1/3 - 1/4 + 1/5 - \ldots$ converges conditionally.

$1 + 1/2 + 1/3 + 1/4 + 1/5 \ldots$ diverges.

• In an *alternating series* in which the signs of the terms alternate positive and negative:

$a_1 + a_2 - a_3 + a_4 - a_5 + a_6 - a_7 + \ldots a_n + \ldots$

The alternating series will converge if the following conditions are true from some point in the series:

$a_n \geq a_{n+1}$ for all values of n, and each a is positive

$\text{Lim}_{n—\infty}\, a_n = 0$

Integral Test for Convergence

Please refer to Chapter 7, "Introduction to the Integral," for an explanation of integrals.

• The *Integral Test* can be applied to a decreasing series of positive terms in which $a_{n+1} < a_n$ for all successive terms. To apply the Integral Test to a series, integrate the function representing the series. If the integral of the series exists and therefore converges, then the series also converges.

Consider the decreasing series:

$$\sum_{n=1}^{m} a_n$$

a_n represents f(x).

If f(x) is a positive continuous function and:

$\int_1^{\infty} f(x)\,dx$ exists and converges then:

$\sum_{n=1}^{m} a_n$ also converges.

• For example, apply the Integral Test to the series represented by f(x) = 1/x.

Integrate between 1 and ∞.

$$\int_1^{\infty} (1/x)dx = \ln x \,\Big|_1^{\infty} = \ln \infty - \ln 1 = \infty$$

The integral of 1/x is ln x, ("ln" is the natural logarithm).

Because the integral from 1 to ∞ is infinity and does not exist, it diverges. Therefore, the series also diverges.

4.7 The Power Series

- The power series of x has the form:

$$a_0 + a_1x + a_2x^2 + a_3x^3 + a_4x^4 + a_5x^5 + a_6x^6 + \ldots a_nx^n + \ldots =$$

$$\sum_{n=0}^{\infty} a_n x^n$$

a_n represents the constant coefficients.

x is the variable.

- A more generalized form of the power series is of $(x - a)$:

$$a_0 + a_1(x-a) + a_2(x-a)^2 + a_3(x-a)^3 + a_4(x-a)^4 + a_5(x-a)^5 + \ldots a_n(x-a)^n + \ldots =$$

$$\sum_{n=0}^{\infty} a_n (x - a)^n$$

- For the power series:

The series in x converges if $x = 0$.

The series in $(x - a)$ converges if $x = a$.

There is a positive number r such that the series converges if $|x| < r$ and diverges if $|x| > r$.

• To evaluate convergence for other values of x for a given power series, the generalized Ratio Test can be applied (see previous section). For example, for the power series in x:

$$\text{Lim}_{n\to\infty} \frac{|a_{n+1}x^{n+1}|}{|a_n x^n|} = \text{Lim}_{n\to\infty} |x| \frac{|a_{n+1}|}{|a_n|}$$

The power series converges when $|x| < \text{Lim}_{n\to\infty} \frac{|a_{n+1}|}{|a_n|}$

and diverges when $|x| > \text{Lim}_{n\to\infty} \frac{|a_{n+1}|}{|a_n|}$

and may or may not converge when

$$|x| = \text{Lim}_{n\to\infty} \frac{|a_{n+1}|}{|a_n|}$$

Note that the set of values of x for which the series is convergent is called the *interval of convergence.*

(See Chapter 5, “Limits,” for explanation of $\text{Lim}_{n\to\infty}$.)

4.8 Expanding Functions into Series

• When a *function* is written in the form of an *infinite series,* it is said to be “expanded” in an infinite series. This series represents all values of x in the *interval of convergence.*

- For the function f(x), the infinite series is:

$$f(x) = \sum_{n=0}^{\infty} a_n (x - a)^n =$$

$$a_0 + a_1(x-a) + a_2(x-a)^2 + a_3(x-a)^3 + a_4(x-a)^4 + a_5(x-a)^5 + \ldots a_n(x-a)^n + \ldots$$

Or:

$$f(x) = \sum_{n=0}^{\infty} a_n x^n =$$

$$a_0 + a_1x + a_2x^2 + a_3x^3 + a_4x^4 + a_5x^5 + a_6x^6 \ldots a_nx^n + \ldots$$

- The function f(x) has the following properties of a polynomial:

  It is continuous within the interval of convergence (there is no break in its graph).

  In series form, the function can be added, subtracted, multiplied, or divided term by term.

  If f(x) is differentiable, then the series can be differentiated term by term (see Chapter 6, "Introduction to the Derivative").

- Two common series representing expansions are the Maclaurin Series and the Taylor Series. In these series, successive derivatives are taken and the coefficients can be obtained. For a function f(x) expanded in a power series:

$$f(x) = \sum_{n=0}^{\infty} a_n (x - a)^n =$$

$$a_0 + a_1(x-a) + a_2(x-a)^2 + a_3(x-a)^3 + \ldots a_n(x-a)^n + \ldots$$

In the special case of $a = 0$:

$f(x) = a_0 + a_1x + a_2x^2 + a_3x^3 + \ldots a_nx^n + \ldots$

Where, $f(0) = a_0$.

Take the first derivative of each term (where ' represents the derivative). (See Chapter 6, "Introduction to the Derivative.")

$f'(x) = a_1 + 2a_2x^1 + 3a_3x^2 + \ldots na_nx^{n-1} + \ldots$

Where, $f'(0) = a_1$.

Take the second derivative of each term.

$f''(x) = 2a_2 + (2)3a_3x + \ldots n(n-1)a_nx^{n-2} + \ldots$

Where, $f''(0) = 2a_2$.

Take the third derivative of each term.

$f'''(x) = (2)3a_3 + (2)(3)4a_4x\ldots n(n-1)(n-2)a_nx^{n-3} + \ldots$

Where $f'''(0) = 6a_3$.

Take the nth derivative of each term.

$f^{(n)}(x) = n!a_n + (n+1)!a_{n+1}x + \ldots$

(Remember "!" represents factorial.)

The coefficients are determined at x = a = 0:

$a_0 = f(0)$

$a_1 = f'(0)$

$a_2 = f''(0)/2$

$a_3 = f'''(0)/(2)(3)$

$a_n = f^{(n)}(0)/n!$

Therefore, the expansion of f(x) is:

$$f(x) = f(0) + f'(0)x + \frac{f''(0)}{2!}x^2 + \frac{f'''(0)}{3!}x^3 + \cdots + \frac{f^{(n)}(0)}{n!}x^n + \cdots$$

$$= a_0 + a_1x + a_2x^2 + a_3x^3 + \ldots a_nx^n + \ldots$$

This is known as the *Maclaurin Series,* which is expanded about the point zero.

- In the *Taylor Series,* the function is expanded about point *a* rather than zero. f(x) is represented as:

$$f(x) = a_0 + a_1(x-a) + a_2(x-a)^2 + a_3(x-a)^3 + \ldots a_n(x-a)^n + \ldots$$

The coefficients a_n are computed by repeated differentiation as with the Maclaurin Series. The expansion of f(x) is:

$$f(x) =$$

$$f(a) + f'(a)(x-a) + \frac{f''(a)}{2!}(x-a)^2 + \frac{f'''(a)}{3!}(x-a)^3 \cdots + \frac{f^{(n)}(a)}{n!}(x-a)^n + \cdots$$

This is known as the Taylor Series, which is expanded about point a. When a = 0, the Taylor Series can be reduced the Maclaurin Series.

• Tables of exponential, logarithmic, and trigonometric functions are often obtained from computations by series. For example, the exponential function e^x can be computed using the Maclaurin expansion.

$$e^x = 1 + x + \frac{x^2}{2!} + \frac{x^3}{3!} + \frac{x^4}{4!} + \cdots + \frac{x^n}{n!} + \cdots$$

for x = 1,

$$e^x = 1 + 1 + \frac{1}{2!} + \frac{1}{3!} + \frac{1}{4!} + \cdots + \frac{1}{n!} + \cdots$$

$= 1 + 1 + 0.5 + 0.166667 + 0.041667 + 0.008333 + 0.001389 + 0.000198 + \ldots = 2.718254$

Therefore, e^x for x = 1 is approximately equal to 2.718254.

Also, trigonometric functions can be expanded and computed for selected values. For example:

$$\sin x = x - \frac{x^3}{3!} + \frac{x^5}{5!} - \cdots + (-1)^{n-1}\frac{x^{2n-1}}{(2n-1)!} + \cdots$$

$$\cos x = 1 - \frac{x^2}{2!} + \frac{x^4}{4!} - \cdots + (-1)^{n-1}\frac{x^{2n-2}}{(2n-2)!} + \cdots$$

4.9 The Binomial Expansion

• The binomial expression (a + b), can be expanded into polynomial form called the *binomial expansion.* (See Chapter 5, "Polynomials," in the second book of the *Master Math* series, *Algebra.*)

• To expand (a + b) into $(a + b)^n$:

First, consider the expansions for $(a + b)^2$, $(a + b)^3$, and $(a + b)^4$:

$$(a + b)^2$$

$$= (a + b)(a + b) = a^2 + ab + ab + b^2 = a^2 + 2ab + b^2$$

$$(a + b)^3 = (a + b)(a + b)(a + b) = a^3 + 3a^2b + 3ab^2 + b^3$$

$$(a + b)^4 = (a + b)(a + b)(a + b)(a + b) = a^4 + 4a^3b + 6a^2b^2 + 4ab^3 + b^4$$

(These expansions are obtained by multiplying the first two binomials, then multiplying each successive binomial with the preceding polynomial.)

For the expansion of $(a + b)^n$, where n is a positive integer, the Binomial Theorem is applied as follows:

$$(a + b)^n$$

$$= a^n + na^{n-1}b + \frac{n(n-1)}{2 \times 1}a^{n-2}b^2 + \frac{n(n-1)(n-2)}{3 \times 2 \times 1}a^{n-3}b^3 + \cdots$$

$$+ \frac{n(n-1)(n-2)\ldots(n-r+2)}{(r-1)!}(a^{n-r+1}b^{r-1}) + \cdots + b^n$$

The rth term is given by:

$$\frac{n(n-1)(n-2)\ldots(n-r+2)}{(r-1)!}(a^{n-r+1}b^{r-1})$$

Chapter 5

Limits

5.1 Introduction to Limits

• In this section, the limit is described, examples of limits are given, as well as rules governing limits of functions that are added, subtracted, multiplied, divided, and raised to a power.

• The limit is used in the estimation of infinite sums and the definition of the derivative in calculus.

• In general, the limit is used to describe *closeness* of a function to a value when the exact value cannot be identified.

- If the $\text{Lim}_{x \to a} f(x) = L$, then it is said that the function f(x) gets close to and may equal some number L as x approaches and gets close to some number *a*.

- For example, consider the function:

$$f(x) = x^2 + 2$$

What will the value of f(x) be close to when x is close to 1?

$$\text{Lim}_{x \to 1} f(x) = \text{Lim}_{x \to 1}(x^2 + 2)$$

For x = 0.9, $f(0.9) = (0.9)^2 + 2 = 2.8$

For x = 1.1, $f(1.1) = (1.1)^2 + 2 = 3.2$

For x = 0.99, $f(0.99) = (0.99)^2 + 2 = 2.98$

For x = 1.01, $f(1.01) = (1.01)^2 + 2 = 3.02$

As x gets closer to 1, $\text{Lim}_{x \to 1}$, then x^2 gets closer to 1 and f(x) gets closer to 3.

$$\text{Lim}_{x \to 1} f(x) = \text{Lim}_{x \to 1}(x^2 + 2) = 3$$

- Example: Consider the limit:

$$\text{Lim}_{x->2}\left(\frac{(x^2 - 4)}{(x - 2)}\right)$$

The limit cannot be found by substituting into the equation. For x = 2:

$$\frac{(x^2 - 4)}{(x - 2)} = 0/0$$

This form of the equation does not enable us to determine if a limit exists.

Instead, by factoring, it is possible to determine if the limit exists. (See Section 5.7 in Chapter 5, "Polynomials," in the second book of the *Master Math* series, *Algebra* for a review of factoring polynomials.)

$$\frac{(x^2 - 4)}{(x - 2)} = \frac{(x + 2)(x - 2)}{(x - 2)} = x + 2$$

Taking the limit:

$$\text{Lim}_{x \to 2}(x + 2) = 4$$

Therefore, as x gets close to 2, $x + 2$ gets close to 4.

• Example: Consider the limit as y gets close to zero for the function:

$$f(y) = 1/y$$

$$\text{Lim}_{y \to 0}(1/y)$$

Because 1/0 is undefined, what will happen to f(y) as y gets close to 0?

If $y = 0.1$, $f(y) = 1/0.1 = 10$

If $y = 0.01$, $f(y) = 1/0.01 = 100$

If $y = 0.001$, $f(y) = 1/0.001 = 1{,}000$

If $y = -0.01$, $f(y) = 1/-0.01 = -100$

If $y = -0.001$, $f(y) = 1/-0.001 = -1{,}000$

As y gets close to 0, 1/y does *not* get closer to any number. The magnitude of 1/y actually increases.

Therefore, $\text{Lim}_{y\to 0}(1/y)$ approaches ∞, and the limit does not exist.

- Example: Consider the limit as x gets close to infinity for the function:

$$f(x) = 1/x$$

$$\text{Lim}_{x\to\infty}(1/x)$$

As x gets very large what happens to 1/x?

If $x = 1{,}000{,}000$,
then $f(x) = 1/1{,}000{,}000 = 0.000001 = 10^{-6}$

If $x = 1{,}000{,}000{,}000$,
then $f(x) = 1/1{,}000{,}000{,}000 = 0.000000001 = 10^{-9}$

If $x = 10^{20}$,
then $f(x) = 1/10^{20} = 0.00000000000000000001 = 10^{-20}$

Therefore, as x becomes large and approaches $\pm\infty$, 1/x becomes very small and approaches zero.

$$\text{Lim}_{x\to\infty}(1/x) = 0$$

- If the limits of the functions f(x) and g(x) exist, then the rules governing limits of functions that are added, subtracted, multiplied, divided, and raised to a power are:

$Lim_{x \to a}(f(x) + g(x)) = Lim_{x \to a}f(x) + Lim_{x \to a}g(x)$

$Lim_{x \to a}(f(x) - g(x)) = Lim_{x \to a}f(x) - Lim_{x \to a}g(x)$

$Lim_{x \to a}(f(x) \times g(x)) = Lim_{x \to a}f(x) \times Lim_{x \to a}g(x)$

$Lim_{x \to a}(f(x) \div g(x)) = Lim_{x \to a}f(x) \div Lim_{x \to a}g(x)$

provided that $Lim_{x \to a}g(x) \neq 0$

$Lim_{x \to a}(f(x))^y = (Lim_{x \to a}f(x))^y$

5.2 Limits and Continuity

- The limit is used to determine if functions are continuous functions or discontinuous functions. Whether or not the graph of a given function is a smooth and continuous curve or line, or whether there are breaks or holes present can be determined using the limit.

- A function is considered continuous at $x = a$, if $Lim_{x \to a}$ exists, $Lim_{x \to a} = f(a)$ and $f(a)$ is defined.

- Whether a function is continuous over the part of the curve near a chosen point, can be determined using the following method. To evaluate whether a function is continuous:

  1. Choose a pair of numbers in the domain set and corresponding range set, relating to the X-axis and Y-axis respectively of the graph.

  2. Take the limit as x gets close to the chosen value in the domain set by choosing x values that are close to x and substituting them into the function.

3. If the function is continuous, the resulting f(x) values in the range set will be close to the chosen f(x) value.

- For example, consider the function:

  $f(x) = 1 + x$

To graph, choose values for x and calculate the corresponding f(x) values.

Values for x are 0, 1, −1, 2

Resulting in f(x) values 1, 2, 0, 3

Resulting in pairs (0, 1), (1, 2), (−1, 0), (2, 3)

Graphing the pairs is depicted as:

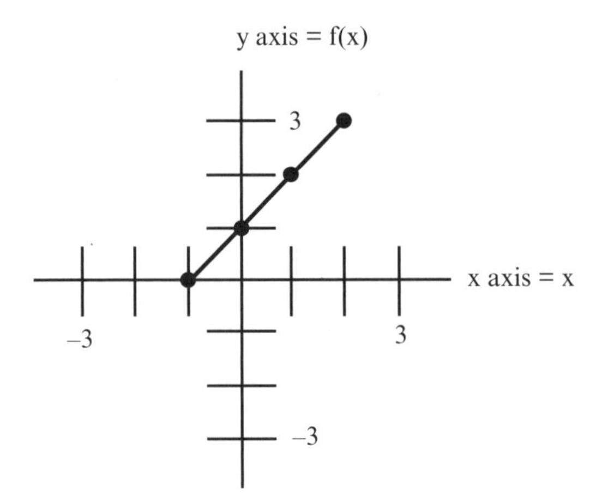

Is this function continuous at point (1, 2)?
(1 is in the domain set and 2 is in the range set.)

Take the limit as x gets close to 1 and determine whether f(x) gets close to 2.

$\text{Lim}_{x \to 1}(1 + x)$

If $x = 1.10$, then $f(1.10) = 1 + 1.10 = 2.10$

If $x = 1.01$, then $f(1.01) = 1 + 1.01 = 2.01$

If $x = 0.90$, then $f(0.90) = 1 + 0.90 = 1.90$

If $x = 0.99$, then $f(0.99) = 1 + 0.99 = 1.99$

Yes, as x gets closer to 1, f(x) gets closer to 2.
Therefore, the function is continuous at this point.

- A method used to visualize whether a function is continuous involves use of symbols such as Epsilon ε and Delta δ to define regions in question on the X and Y axes of the graph of a function. Consider the limit of the function f(x).

$$\text{Lim}_{x \to a} f(x) = L$$

Where:

1. The limit exists.
2. ε represents an error tolerance allowed for L.
3. δ represents the distance that x is from $x = a$.

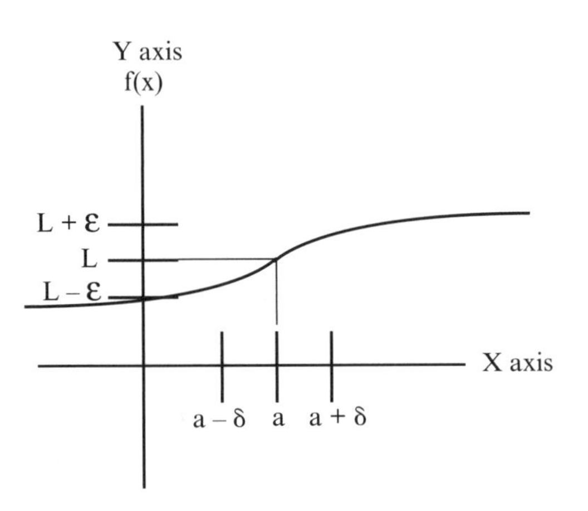

In this method, a value for ε is chosen and $\delta(\varepsilon)$ is the result. (Or a value for δ is chosen and $\varepsilon(\delta)$ is the result.)

If $\text{Lim}_{x \to a} f(x) = f(a) = L$ and the limit exists, such that:

$$L + \varepsilon = f(a) + \varepsilon$$
$$L = f(a)$$
$$L - \varepsilon = f(a) - \varepsilon$$

Then the following is true:

$f(a) + \varepsilon > f(x) > f(a) - \varepsilon$ or equivalently, $|f(x) - L| < \varepsilon$

For every chosen number ε where $\varepsilon > 0$, there is a positive number for $\delta(\varepsilon)$ that results.

For a chosen f(x), x must be within $a - \delta$ and $a + \delta$, such that:

$a - \delta < x < a + \delta$ or equivalently, $0 < |x - a| < \delta$

Also, f(x) is a continuous function at x = a if: f is defined on the open interval containing "a", the $\text{Lim}_{x \to a} f(x) = f(a)$, and the limit exists.

- A *polynomial function*, in general, is continuous everywhere and its graph is a continuous curve. However, a polynomial function is not continuous if there are ratios of polynomial functions and the denominator is zero. The point where a denominator is zero, the function is discontinuous.

- A quadratic function is a type of polynomial function and is continuous everywhere. Taking the limit:

$$\text{Lim}_{x \to a} f(x) = \text{Lim}_{x \to a}(Ax^2 + Bx + C) = f(a)$$

- A function that is not continuous may be discontinuous at a single point. For example, the following function is continuous except at a=0 where it is discontinuous: (Remember 1/0 is undefined.)

$$\text{Lim}_{x \to a} f(x) = \text{Lim}_{x \to a} 1/x = 1/a$$

- The following is a graph of a function that is discontinuous at a point:

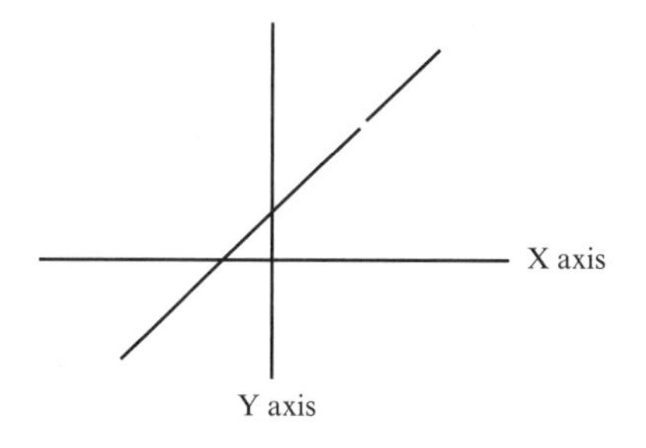

One point can be inserted to make the graph continuous.

• The following is a graph of a function that is discontinuous at more than one point:

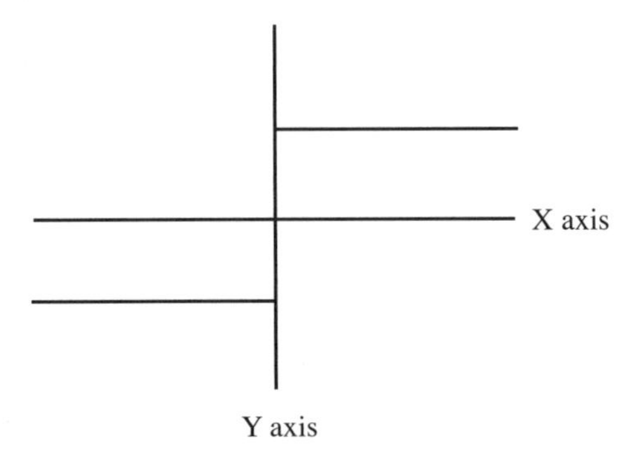

One point cannot be inserted to make the graph continuous; it has a "jump."

• In summary:

If $\text{Lim}_{x\to a}f(x)$ exists and is equal to L, but f(x) does not exist when $x = a$ so that $f(a) \neq L$, then the graph of f(x) is discontinuous at the point $x = a$. In this case there is only discontinuity at a single point.

If $\text{Lim}_{x\to a}f(x)$ does not exist because as x approaches a from either $x>a$ or $x<a$, the value of f(x) approached from $x>a$ is different from the value of f(x) approached from $x<a$, then the graph of f(x) is discontinuous and has a jump at point $a = x$.

If $\text{Lim}_{x\to a}f(x)$ does not exist because as x approaches a, the absolute value of f(x), $|f(x)|$, gets larger and larger, then the graph is "infinitely discontinuous" at $x = a$.

Chapter 6

Introduction to the Derivative

6.1 Definition

• The *derivative* is used to describe the rate of change of something, such as velocity, as well as the concept of the tangent to a curve (see Section 6.5). There are many applications of the derivative including but not limited to tangents, slopes, rates of change, curvilinear and straight-line motion, maxima and minima, and tests for extrema.

• For example, the time rate of change of an object in motion such as a car, plane, pitcher's fast ball, etc., is the rate of change of distance with respect to time, and is called *velocity*. In other words, velocity is the derivative or rate of change of distance with respect to time.

• Remember, from rate problems, distance equals rate times time, or d = rt. (d = rt was first introduced in Section 1.4, "Simple word problems," in the second book of the *Math Masters* series, *Algebra*.) Therefore:

$$\text{Rate} = \frac{\text{Distance}}{\text{Time}}$$

The time rate of change of distance is velocity.

$$\text{Average Velocity} = \frac{\text{Change in Distance}}{\text{Change in Time}}$$

Also, the time rate of change of velocity is *acceleration.*

$$\text{Acceleration} = \frac{\text{Change in Velocity}}{\text{Change in Time}}$$

• The derivative can be used to study the rate of change of something that changes or varies with respect to some parameter such as position or time. For example:

The number of ants in an ant farm population over a period of time.

The rate of change in the cost of producing computers with respect to the number of computers manufactured. (Producing higher volumes costs less because parts are purchased at volume discounts.)

• To develop the *definition of the derivative*, consider the velocity of an airplane flying from the east coast to the west coast.

The distance the airplane is from its starting point or any defined reference point is a function of time (depends on time) or f(t). (In this example, f is the distance function.)

At time = t, the airplane is f(t) units from the starting or reference point. (The units could be hours.)

At time = t + h, the airplane is f(t + h) units from the starting or reference point and *h represents an increment of time.*

The change in the position of the airplane during the increment of time h is f(t+h) – f(t).

The rate of change of the distance with respect to time between time = t and time = t+h is the average velocity of the airplane.

The average velocity during this time period =

$$\frac{(\text{Position at time} = t + h) - (\text{Position at time} = t)}{\text{Increment of time}}$$

$$= \frac{f(t + h) - f(t)}{h}$$

To find the velocity of the airplane at a particular point when time = t, shrink the time increment h surrounding time t. The velocity at the point where time = t is called the *instantaneous velocity*, and is determined by taking the limit as the increment of time h shrinks to zero.

$$\text{Velocity at time } t = v(t) = \text{Lim}_{h \to 0} \frac{f(t + h) - f(t)}{h}$$

As h gets close to zero (but not equal to zero), the time increment h and the distance $f(t+h) - f(t)$ will get smaller.

- To determine velocity at time t or v(t):

  1. Determine $f(t+h)$ and $f(t)$.
  2. Subtract $f(t)$ from $f(t+h)$.
  3. Divide $f(t+h) - f(t)$ by h.
  4. Take the limit as h approaches zero.

- Example: Find the velocity at t = 2 hours, if the distance in miles is represented by $f(t) = 3t^2$.

Determine f(t+h) and f(t).

$$f(t+h) = 3(2 + h)^2 = 3(2 + h)(2 + h)$$

$$= 3(4 + 2h + 2h + h^2) = 3(4 + 4h + h^2)$$

$$= 12 + 12h + 3h^2$$

$$f(t) = 3(2)^2 = 12$$

Subtract f(t) from f(t+h).

$$f(t+h) - f(t) = (12 + 12h + 3h^2) - 12 = 12h + 3h^2$$

Divide f(t+h) – f(t) by h.

$$\frac{12h + 3h^2}{h} = \frac{12h}{h} + \frac{3h^2}{h} = 12 + 3h$$

Take the limit as h approaches zero.

$$v(2hrs) = \mathrm{Lim}_{h\to 0}(12 + 3h) = 12$$

Therefore, the velocity at t = 2 hours is 12 miles/hour.

- The definition of the derivative with respect to time of the distance function f(t) can be written:

$$\frac{df(t)}{dt} = \mathrm{Lim}_{h\to 0} \frac{f(t+h) - f(t)}{h}$$

provided the limit exists.

- The definition of the derivative with respect to some function of x can be written:

$$\frac{df(x)}{dx} = \text{Lim}_{h \to 0} \frac{f(x+h) - f(x)}{h}$$

- Symbols for the derivative of a function f include:

$$\frac{df}{dx}, \left(\frac{d}{dx}\right)f, f', \mathbf{D}f$$

- Symbols for the derivative of a function of f(x) include:

$$\frac{df(x)}{dx}, \left(\frac{d}{dx}\right)f(x), f'(x), \mathbf{D}f(x), \mathbf{D}_x f(x)$$

- Because velocity is the derivative of the distance function, the velocity at time t describes the distance at f(t).

$$v(t) = \frac{df(t)}{dt}$$

If x = distance and x(t) = distance with respect to time, then:

$$v(t) = x'(t) = \frac{dx(t)}{dt}$$

- Acceleration is the time rate of change of velocity, and acceleration depends on time.

$$\text{Acceleration} = a(t) = v'(t) = x''(t)$$

• To evaluate a derivative, a formula described in the following section is generally used rather than the definition of the derivative.

6.2 Evaluating Derivatives

• Evaluating derivatives using the definition of the derivative is labor intensive.

$$\frac{df(x)}{dx} = \text{Lim}_{h \to 0} \frac{f(x+h) - f(x)}{h}$$

Instead, there is a shortcut formula used to evaluate derivatives. This *derivative formula* is derived from the definition of the derivative (see *Master Math: Calculus* for the derivation). The derivative formula can be applied as follows:

$$\frac{d}{dx} x^n = nx^{n-1}$$

Where x represents any variable and n is any number.

$$\frac{d}{dx} cx^n = cnx^{n-1}$$

Where c represents any constant number.

$$\frac{d}{dx} c = 0$$

The derivative of a constant number by itself is zero.

$$\frac{d}{dx}cf(x) = c\frac{d}{dx}f(x)$$

Where f(x) represents any function.

- The following are examples of using the derivative formula to evaluate derivatives:

$$\frac{d}{dx}x = 1 \times x^{1-1} = 1x^0 = 1$$

$$\frac{d}{dx}x^2 = 2 \times x^{2-1} = 2x^1 = 2x$$

$$\frac{d}{dx}x^3 = 3 \times x^{3-1} = 3x^2$$

$$\frac{d}{dx}x^4 = 4 \times x^{4-1} = 4x^3$$

$$\frac{d}{dx}x^{25} = 25 \times x^{25-1} = 25x^{24}$$

$$\frac{d}{dx}\sqrt{x} = \frac{d}{dx}x^{1/2} = (1/2) \times x^{(1/2)-1} = (1/2) \times x^{(1/2)-(2/2)}$$

$$= (1/2)x^{-1/2} = 1/(2x^{1/2}) = 1/(2\sqrt{x})$$

(Note: $x^n = 1/x^{-n}$ and $x^{-n} = 1/x^n$.)

$$\frac{d}{dx}(1/x) = \frac{d}{dx}x^{-1} = -1 \times x^{-1-1} = -1x^{-2} = -1/x^2$$

$$\frac{d}{dx}(1/x^4) = \frac{d}{dx}x^{-4} = -4 \times x^{-4-1} = -4x^{-5} = -4/x^5$$

$$\frac{d}{dx}(2x) = 2x^{1-1} = 2x^0 = 2$$

$$\frac{d}{dx}(2x^2) = 2 \times 2x^{2-1} = 2 \times 2x^1 = 4x$$

$$\frac{d}{dx}2 = 0$$

6.3 Differentiating Multivariable Functions

• The derivative formula is used to differentiate multivariable functions with respect to one of the variables.

• When a function that contains more than one variable is differentiated, the derivative formula is applied to the variable that is being differentiated.

• Apply the derivative formula to the variable (x, y, or z) in each multivariable function that it is differentiated with respect to:

$\frac{d}{dx}(x^2y^2) = 2xy^2$ Differentiated with respect to x.

$\frac{d}{dy}(x^2y^2) = 2x^2y$ Differentiated with respect to y.

$\frac{d}{dz}(x^2y^2z^2) = 2x^2y^2z$ Differentiated with respect to z.

6.4 Differentiating Polynomials

• To differentiate a polynomial function, the derivative formula is applied to each term separately. Remember, each term is separated by a plus or minus sign.

• For example, differentiate the following polynomial function term by term using the derivative formula:

$$\frac{d}{dx}(x^3 + x^2 + x + c)$$

$$= \frac{d}{dx}x^3 + \frac{d}{dx}x^2 + \frac{d}{dx}x + \frac{d}{dx}c$$

$$= 3x^2 + 2x + 1 + 0$$

6.5 Derivatives and Graphs of Functions

• In this section, the relationship between graphs of functions and the derivative, the slope, and the tangent is discussed.

• In the graph of a function, the slope of a line drawn tangent to the curve through point (a, f(a)) on the curve is the derivative of the function at point (a, f(a)).

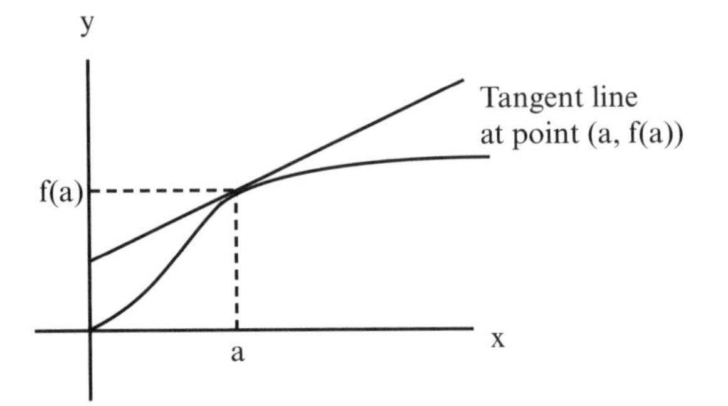

The *slope of the tangent* at point (a, f(a)) equals the derivative f'(a). (Note that if a tangent line is vertical, its slope is undefined.)

• The definition of the derivative can be used to prove that the slope of a line drawn tangent to a graph of a function is the derivative at that point.

Imagine two points on the curve, (a, f(a)) and (a+h, f(a+h)), with a tangent line drawn through each point. Tangent 1 is drawn through point (a, f(a)) and Tangent 2 is drawn through point (a+h, f(a+h)).

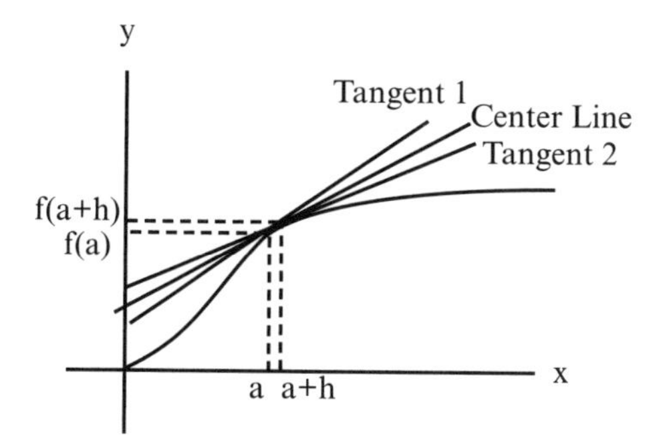

If a third line is drawn through the two tangent points (center line shown in illustration), the slope of the center line is given by the change in y over the change in x, between the two points.

Therefore, the slope of the line through (a, f(a)) and (a+h, f(a+h)) is given by:

$$\frac{f(a+h)-f(a)}{a+h-a}=\frac{f(a+h)-f(a)}{h}$$

The slope of this line is approximately equal to the slope of Tangent 1. If the value of the increment h between the two points is reduced, the value of h will approach zero, and the Tangent 2 line through point (a+h, f(a+h)) will approach being equal to the Tangent 1 line through point (a, f(a)).

Also, the line drawn through both points will approach becoming equal to the Tangent 1 and Tangent 2 lines. Therefore, the following is true:

$$\text{Lim}_{h \to 0} \frac{f(a+h) - f(a)}{h} = \text{the slope of Tangent 1},$$

provided the limit exists.

This is the definition of the derivative and equals f'(a).

Therefore, the slope of the tangent at (a, f(a)) is the derivative of f at a.

- The equation for a line passing through some point (a, b) with a slope of m is given by:

$$y - b = m(x - a)$$

Because the tangent line passes through point (a, f(a)) and the slope is f'(a), the equation for the tangent line is written:

$$y - f(a) = f'(a)(x - a)$$

or by rearranging,

$$f'(a) = \frac{y - f(a)}{x - a}$$

- In summary:

$$\frac{dy}{dx} = y' = m = \text{slope of the tangent to a curve } y = f(x)$$

- Note: If for a given point on the graph of a function, the derivative does not exist, then that point may be at the end of the curve, at a corner on the curve, or the tangent line is a vertical line and has no slope. Also, if the derivative does not exist at a point, then the graph may be discontinuous at that point. Conversely, if the derivative does exist at a point, then the graph is continuous at that point.

6.6 Adding and Subtracting Derivatives of Functions

- To differentiate functions that are added or subtracted, differentiate each function separately, then add or subtract. Also, if one of the functions contains a polynomial, remember to differentiate the polynomial term by term.

- Differentiate the sum of the two functions f(x) and g(x).

$$\frac{d}{dx}[f(x) + g(x)] = \frac{d}{dx}f(x) + \frac{d}{dx}g(x)$$

$$\frac{d}{dx}[f(x) - g(x)] = \frac{d}{dx}f(x) - \frac{d}{dx}g(x)$$

(See *Master Math: Calculus* for proofs using the definition of the derivative.)

- For example:

If $f(x) = 2x^2$ and $g(x) = x^3 + 3$, find $\frac{d}{dx}[f(x) + g(x)]$.

$$\frac{d}{dx}[f(x) + g(x)] = \frac{d}{dx}[(2x^2) + (x^3 + 3)]$$

$$= \frac{d}{dx}(2x^2) + \frac{d}{dx}(x^3 + 3) = \frac{d}{dx}2x^2 + \frac{d}{dx}x^3 + \frac{d}{dx}3$$

$$= 4x + 3x^2 + 0 = 4x + 3x^2$$

6.7 Multiple or Repeated Derivatives of a Function

- To find the second, third, fourth, etc., derivative of a function, repeatedly apply the derivative formula for each specified derivative.

- Example: What is the second derivative of $f(x) = x^2$?

$f(x) = x^2$

Take first derivative.

$f'(x) = 2x$

Take second derivative.

$f''(x) = 2$

• Notation for taking multiple derivatives is:

For the nth derivative:

$$\frac{d^n f(x)}{dx^n},\ \left(\frac{d^n}{dx^n}\right) f(x),\ f^n(x),\ \mathbf{D}^n f(x),\ \mathbf{D}_x^n f(x)$$

For the second derivative:

$$\frac{d^2 f(x)}{dx^2},\ \left(\frac{d^2}{dx^2}\right) f(x),\ f^2(x),\ f''(x),\ \mathbf{D}^2 f(x),\ \mathbf{D}_x^2 f(x)$$

For the third derivative:

$$\frac{d^3 f(x)}{dx^3},\ \left(\frac{d^3}{dx^3}\right) f(x),\ f^3(x),\ f'''(x),\ \mathbf{D}^3 f(x),\ \mathbf{D}_x^3 f(x)$$

6.8 Derivatives of Products and Powers of Functions

• In this section, differentiating products and powers of functions using the product rule, extensions of the product rule, and an alternative method are described.

• The *product rule* can be used to differentiate the product of two functions. The product rule applied to the product of the two functions f(x) and g(x) is:

$$\frac{d}{dx}[f(x) \times g(x)] = [\frac{d}{dx} f(x)] \times g(x) + f(x) \times [\frac{d}{dx} g(x)]$$

Using shorthand notation the product rule is written:

$$(fg)' = f'g + fg'$$

(See *Master Math: Calculus* for a proof of the product rule.)

- Example: If $f(x) = x^2$ and $g(x) = x^3 + 3$, find

$\frac{d}{dx}[f(x) \times g(x)]$ using the product rule.

To differentiate the product $\frac{d}{dx}[(x^2) \times (x^3 + 3)]$ using the product rule, $(fg)' = f'g + fg'$:

First evaluate f' and g'.

$$f'(x) = 2x$$

$$g'(x) = 3x^2 + 0 = 3x^2$$

Apply the product rule.

$$\frac{d}{dx}[(x^2) \times (x^3 + 3)]$$

$$= [\frac{d}{dx}(x^2)] \times (x^3 + 3) + (x^2) \times [\frac{d}{dx}(x^3 + 3)]$$

$$= [2x] \times (x^3 + 3) + (x^2) \times [3x^2]$$

Multiply using the distributive property.

$$= (2x^4 + 6x) + (3x^4)$$

Combine like terms.

$$= (2x^4 + 3x^4) + 6x = 5x^4 + 6x$$

Therefore, $\frac{d}{dx}[(x^2) \times (x^3 + 3)] = 5x^4 + 6x$.

(Note that in this example, the monomial and binomial can be multiplied first, then differentiated term by term.)

• An extension of the product rule can be applied to derivatives of multiple products.

For two functions f and g:

$$(fg)' = f'g + fg'$$

For three functions f, g, and h:

$$(fgh)' = f'gh + fg'h + fgh'$$

For four functions f, g, h, and p:

$$(fghp)' = f'ghp + fg'hp + fgh'p + fghp'$$

• The product rule can be applied to find the derivative of a function raised to the second power.

• For example, use the product rule to find the derivative of the binomial $(x^3 + x^2)$ raised to the second power.

$$\frac{d}{dx}f(x) = \frac{d}{dx}[(x^3 + x^2)^2] = \frac{d}{dx}[(x^3 + x^2)(x^3 + x^2)]$$

Apply the product rule.

$$\frac{d}{dx}[f(x) \times g(x) = [\frac{d}{dx}f(x)] \times g(x) + f(x) \times [\frac{d}{dx}g(x)]$$

Substituting: $\frac{d}{dx}[(x^3 + x^2) \times (x^3 + x^2)]$

$$= [\frac{d}{dx}(x^3 + x^2)] \times (x^3 + x^2) + (x^3 + x^2) \times [\frac{d}{dx}(x^3 + x^2)]$$

Differentiate each binomial term by term.

$$= (3x^2 + 2x) \times (x^3 + x^2) + (x^3 + x^2) \times (3x^2 + 2x)$$

Multiply the binomials.

$$= (3x^5 + 3x^4 + 2x^4 + 2x^3) + (3x^5 + 2x^4 + 3x^4 + 2x^3)$$

Combine like terms.

$$= 3x^5 + 3x^5 + 3x^4 + 2x^4 + 2x^4 + 3x^4 + 2x^3 + 2x^3$$

$$= (3 + 3)x^5 + (3 + 2 + 2 + 3)x^4 + (2 + 2)x^3$$

$$= 6x^5 + 10x^4 + 4x^3$$

Therefore, $\frac{d}{dx}[(x^3 + x^2)^2] = 6x^5 + 10x^4 + 4x^3$.

(Note that in this example, the two binomials can be multiplied first, then differentiated term by term.)

• An alternative method can be used to evaluate the derivatives of products and powers of functions.

$$\frac{d}{dx}(f(x))^2 = 2 \times f(x) \times \frac{d}{dx}f(x)$$

$$\frac{d}{dx}(f(x))^3 = 3 \times (f(x))^2 \times \frac{d}{dx}f(x) = 3 \times f(x) \times f(x) \times \frac{d}{dx}f(x)$$

$$\frac{d}{dx}(f(x))^4 = 4 \times (f(x))^3 \times \frac{d}{dx}f(x)$$

$$= 4 \times f(x) \times f(x) \times f(x) \times \frac{d}{dx}f(x)$$

$$\frac{d}{dx}(f(x))^n = n \times (f(x))^{n-1} \times \frac{d}{dx}f(x)$$

• To compare this method with the product rule, evaluate the derivative in the previous example of the binomial function.

$$f(x) = (x^3 + x^2)^2$$

Find f'(x) using $\frac{d}{dx}(f(x))^n = n \times (f(x))^{n-1} \times \frac{d}{dx}f(x)$.

$$\frac{d}{dx}(x^3 + x^2)^2 = 2 \times (x^3 + x^2)^{2-1} \times (\frac{d}{dx}(x^3 + x^2))$$

$$= 2 \times (x^3 + x^2) \times (3x^3 + 2x)$$

Multiply the two binomials.

$$= 2 \times (3x^5 + 2x^4 + 3x^4 + 2x^3)$$

Multiply the 2 using the distributive property.

$$= 6x^5 + 4x^4 + 6x^4 + 4x^3$$

Combine like terms.

$$= 6x^5 + 10x^4 + 4x^3$$

Therefore, $\frac{d}{dx}[(x^3 + x^2)^2] = 6x^5 + 10x^4 + 4x^3$.

This is the same result obtained above using the product rule.

6.9 Derivatives of Quotients of Functions

- The *quotient rule* can be applied to evaluate derivatives of quotients of functions. For the functions f(x) and g(x) the quotient rule is:

$$\frac{d}{dx}\left(\frac{f(x)}{g(x)}\right) = \frac{\frac{d}{dx}(f(x)) \times g(x) - f(x) \times \frac{d}{dx}g(x)}{g(x)^2}$$

Using notation, the quotient rule can be written:

$$\left(\frac{f}{g}\right)' = \frac{f'g - fg'}{g^2}$$

(See *Master Math: Calculus* for a proof of the quotient rule.)

• For example, use the quotient rule to find the derivative of the quotient of functions f and g.

$f(x) = x^2 + 2$ and $g(x) = x^3 + 3$

To find the derivative of the quotient:

$$\frac{d}{dx}\left(\frac{f(x)}{g(x)}\right) = \frac{d}{dx}\left(\frac{x^2+2}{x^3+3}\right)$$

First evaluate $f'(x)$ and $g'(x)$.

$$f'(x) = 2x$$

$$g'(x) = 3x^2$$

Apply the quotient rule:

$$\left(\frac{f}{g}\right)' = \frac{f'g - fg'}{g^2}$$

The quotient becomes:

$$\frac{d}{dx}\left(\frac{x^2+2}{x^3+3}\right)$$

$$= \frac{(\frac{d}{dx}(x^2+2)) \times (x^3+3) - (x^2+2) \times (\frac{d}{dx}(x^3+3))}{(x^3+3)^2}$$

Take derivatives.

$$= \frac{(2x)(x^3+3) - (x^2+2)(3x^2)}{(x^3+3)(x^3+3)}$$

Multiply.

$$= \frac{(2x^4 + 6x) - (3x^4 + 6x^2)}{x^6 + 3x^3 + 3x^3 + 9}$$

Combine like terms.

$$= \frac{2x^4 - 3x^4 - 6x^2 + 6x}{x^6 + 6x^3 + 9}$$

$$= \frac{-x^4 - 6x^2 + 6x}{x^6 + 6x^3 + 9}$$

Therefore, $\frac{d}{dx}\left(\frac{x^2 + 2}{x^3 + 3}\right) = \frac{-x^4 - 6x^2 + 6x}{x^6 + 6x^3 + 9}$.

6.10 The Chain Rule for Differentiating Complicated Functions

- The *chain rule* can be used to differentiate functions in which variables depend on other variables. Consider the function f that depends on the variable u, but u depends on the variable x. In other words, f is a function of u, and u is a function of x. The chain rule is:

$$\frac{df(u(x))}{dx} = \frac{df(u)}{du} \times \frac{du(x)}{dx}$$

- For example, if $f(u) = (u(x))^2$ and $u(x) = x^3$, the chain rule is:

$$\frac{df(u(x))}{dx} = \frac{d(u(x))^2}{du} \times \frac{dx^3}{dx} = 2u(x) \times 3x^2$$

Substitute $u(x) = x^3$.

$$= 2x^3 \times 3x^2 = 6x^5$$

Therefore, $\frac{df(u(x))}{dx} = \frac{d(u(x))^2}{du} \times \frac{dx^3}{dx} = 6x^5$.

Because this is a simple example, $u(x) = x^3$ can be substituted into $f(u) = (u(x))^2$ directly, and the derivative taken.

$$f(u) = (u(x))^2 = (x^3)^2 = x^6$$

$$\frac{df(u(x))}{dx} = \frac{dx^6}{dx} = 6x^5$$

• The chain rule can be used to break complex functions into two simpler functions. Consider the derivative:

$$\frac{d}{dx}f(x) = \frac{d}{dx}[(x^2 + x)^3]$$

To simplify and use the chain rule, let $f(u) = (u(x))^3$ and $u(x) = (x^2 + x)$.

Using the chain rule, $\frac{df(u(x))}{dx} = \frac{df(u)}{du} \times \frac{du(x)}{dx}$,

substitute for $f(u)$ and $u(x)$.

$$\frac{df(u(x))}{dx} = \frac{d(u(x))^3}{du} \times \frac{d(x^2 + x)}{dx}$$

Differentiate.

$$\frac{df(u(x))}{dx} = 3 \times (u(x))^2 \times (2x + 1)$$

Substitute $u(x) = (x^2 + x)$.

$$\frac{df(u(x))}{dx} = 3 \times (x^2 + x)^2 \times (2x + 1)$$

$$= 3 \times (x^2 + x) \times (x^2 + x) \times (2x + 1)$$

Multiply the first two binomials.

$$= 3 \times (x^4 + x^3 + x^3 + x^2) \times (2x + 1)$$

Combine like terms.

$$= 3 \times (x^4 + 2x^3 + x^2) \times (2x + 1)$$

Multiply the binomial and the trinomial.

$$= 3 \times [(2x^5 + 4x^4 + 2x^3) + (x^4 + 2x^3 + x^2)]$$

Combine like terms, then multiply the 3.

$$= 3 \times [2x^5 + 5x^4 + 4x^3 + x^2] = 6x^5 + 15x^4 + 12x^3 + 3x^2$$

Therefore, $\frac{d}{dx}[(x^2 + x)^3] = 6x^5 + 15x^4 + 12x^3 + 3x^2$.

• Note that the method described in Section 6.8, "Derivatives of Products and Powers of Functions," can also be applied to this problem, using:

$$\frac{d}{dx}(f(x))^n = n \times (f(x))^{n-1} \times \frac{d}{dx}f(x)$$

• The chain rule can be applied to the parametric equations. If a point (x, y) can be described in terms of a third variable t, called the parameter, the equations are called parametric equations. If $x = f(t)$ and $y = g(t)$, then dx/dt and dy/dt can be determined using the chain rule.

$$\frac{dy}{dx} = \frac{dy}{dt}\frac{dt}{dx} = \frac{dy/dt}{dx/dt}$$

6.11 Differentiation of Implicit vs. Explicit Functions

- If y is given explicitly as a function of x, it is not difficult to obtain $\frac{dy}{dx}$, because if $y = f(x)$, then $\frac{dy}{dx} = f'(x)$.

This is *explicit differentiation.* However, if y is given implicity as a function of x, for example, $f(x, y) = 2$, then rather than solving the equation for y first, the equation can be differentiated as it is term by term, then solved for $\frac{dy}{dx}$ in terms of x and y. This called *implicit differentiation.*

- Consider a function that is explicitly in terms of x:

$$f(x) = 2x^2 + 4$$

If we let $f(x) = y(x)$, then:

$$y(x) = 2x^2 + 4$$

Rearranging.

$$2x^2 + 4 - y(x) = 0$$

Therefore, y is given implicitly as a function of x.

• For example, evaluate dy/dx for the above relation, which implicitly gives y as a function of x.

$$2x^2 + 4 - y(x) = 0$$

Rearrange.

$$y(x) = 2x^2 + 4$$

Take the derivative of each term with respect to x.

$$\frac{dy(x)}{dx} = \frac{d2x^2}{dx} + \frac{d4}{dx}$$

$$\frac{dy(x)}{dx} = 4x$$

• Example: Suppose we want to find the slope of a line tangent to the graph of a function such as:

$$xy(x) + x = 5$$

Remember, in the graph of a function, the slope of a line drawn tangent to the curve through a point on the curve is the derivative of the function at that point. To find the slope of a line tangent to the graph, the derivative of y with respect to x, or dy/dx, must be taken.

Take the derivative term by term.

$$\frac{d(xy(x))}{dx} + \frac{dx}{dx} = \frac{d5}{dx}$$

Evaluate the first term using the product rule

(fg)' = f'g + fg'.

$$\frac{d(xy(x))}{dx} = y(x) \times \frac{dx}{dx} + x \times \frac{dy(x)}{dx}$$

$$\frac{d(xy(x))}{dx} = y(x) + x \times \frac{dy(x)}{dx}$$

Evaluate the second term.

$$\frac{dx}{dx} = 1$$

Evaluate the third term.

$$\frac{d5}{dx} = 0$$

The entire function becomes:

$$y(x) + x \times \frac{dy(x)}{dx} + 1 = 0$$

Rearrange to isolate dy/dx.

$$x \times \frac{dy(x)}{dx} = -y(x) - 1$$

$$\frac{dy(x)}{dx} = \frac{(-y(x) - 1)}{x}$$

This equation represents the slope of the tangent line at point (x, y(x)).

6.12 Using Derivatives to Determine the Shape of the Graph of a Function (Minimum and Maximum Points)

• Evaluating first and second derivatives of functions to find *minimum* and *maximum* points, and local *extrema* of a function is a common application of the derivative. When experiments or evaluations are conducted in science, business, engineering, etc., data is gathered, relationships are developed, and graphs are constructed in order to assist in the understanding of the data and to predict future patterns and events. Information depicted in the graphs, such as where the graph is rising or falling, convex or concave, and where the high and low points are (which correspond to the maximum and minimum values), are all crucial to the evaluation of the data.

• For example, if a farmer develops and graphs a function to describe the population of bees on her farm during some period of time, she can determine whether the population of bees is increasing or decreasing during certain smaller time intervals by the sign of the derivative at those intervals.

• The sign of the derivative of a function describes the shape of the graph of the function at the point where the derivative is taken. For the graph:

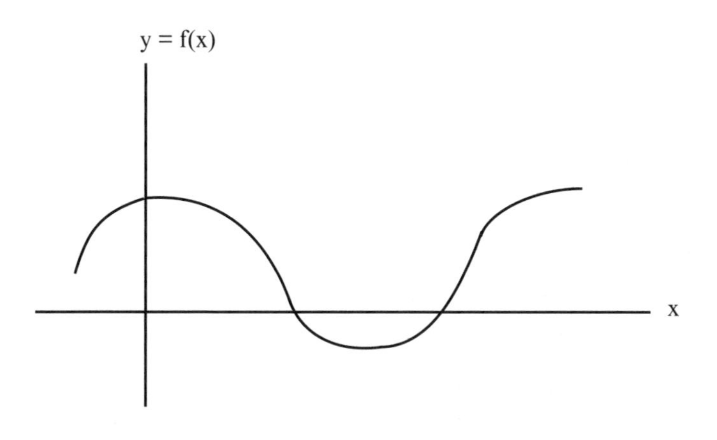

If f(x) is decreasing as x is increasing, the sign of the derivative is negative. Therefore, $f'(x) < 0$ where the graph of f is decreasing.

If f(x) is increasing as x is increasing, the sign of the derivative is positive. Therefore, $f'(x) > 0$ where the graph of f is increasing.

If the graph of the function is horizontal, the derivative of f is zero. Therefore, $f'(x) = 0$ where the graph of f is horizontal.

Minimum and Maximum Points

• A point where the graph of a function is horizontal may represent a minimum or maximum point. There are examples where a graph will not have a minimum or maximum, such as if the graph is a straight horizontal or vertical line. A minimum or maximum on the graph may be the minimum or maximum of the function, or there may be many "local" minimum or maximum points called *local extrema.*

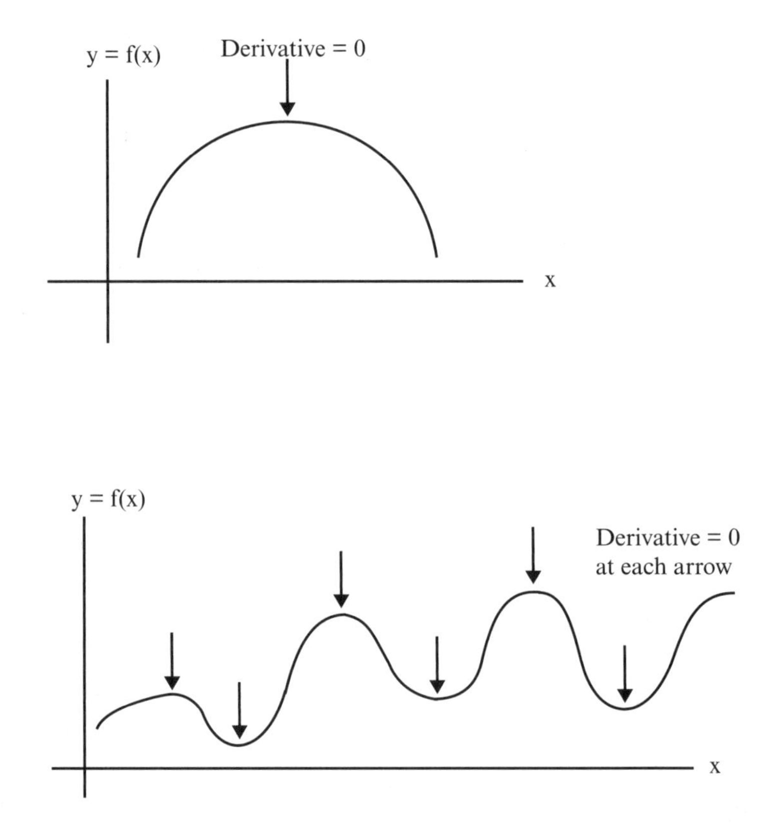

- Consider the following graph of continuous function f:

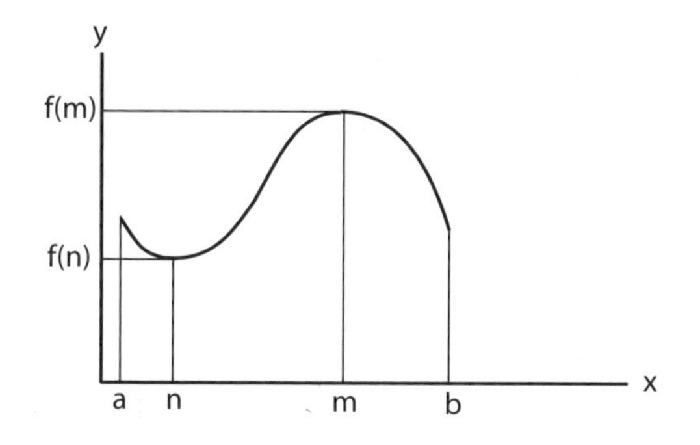

If the highest point on f is the point (m, f(m)), then f(m) is the maximum value of f, and $f(m) \geq f(x)$ for all x. In this graph there are two extrema points, a minimum and a maximum in between points a and b.

- To find the *local extrema* of a function, the graph can be inspected or the derivative can be evaluated. At the extrema points, the derivative of the function f is equal to zero.

If f(n) is a minimum point and f(m) is a maximum point, and:

If $f'(n)$ and $f'(m)$ exist, then $f'(n) = 0$ and $f'(m) = 0$.

- As a general rule, for a given function f, all values of x where $f'(x) = 0$ or where $f'(x)$ is undefined, represent all possible local extrema. Note that there may be cases where $f'(x) = 0$ but an extrema does not exist.

- To find all possible local extrema (minimum or maximum values of f) within some interval between $x=a$ and $x=b$ or between points $(a, f(a))$ and $(b, f(b))$:

    1. Find all x values that satisfy $f'(x) = 0$ or $f'(x) = \text{undefined}$.

    2. Evaluate each x value found in the first step by substituting it into the function f.

    3. Evaluate values of x at the ends of the interval (at a and b) to find $f(a)$ and $f(b)$.

    4. The largest value in the second step is the maximum of $f(x)$ and the smallest value is the minimum of $f(x)$ within the interval $a-b$.

- For example, find the minimum and maximum of the function $f(x) = x^2 + 2x$ between the interval of $x = 0$ and $x = -2$ where $-2 \leq x \leq 0$.

Find all x values that satisfy $f'(x) = 0$ or $f'(x) = \text{undefined}$.

$$f'(x) = \frac{d}{dx}x^2 + \frac{d}{dx}2x = 2x + 2$$

Where does $f'(x) = 0$?

Because $f'(x) = 2x + 2$, set $2x + 2 = 0$,

$2x + 2 = 0$

Solve for x.

$2x = -2$

$x = -2/2 = -1$

Evaluate each x value found by substituting it into the function f. Evaluate f(x) at $x = -1$.

$f(-1) = x^2 + 2x = (-1)^2 + 2(-1) = 1 + -2 = -1$

Evaluate the values of x at the ends of the interval (at a and b) to find f(a) and f(b).

Evaluate $f(x) = x^2 + 2x$ at the end points -2 and 0.

$f(-2) = (-2)^2 + 2(-2) = 4 + -4 = 0$

$f(0) = (0)^2 + 2(0) = 0 + 0 = 0$

Therefore, the number for the critical points of f over this interval ($-2 \leq x \leq 0$) are:

$f(-1) = -1$, $f(0) = 0$, and $f(-2) = 0$

The largest and smallest values from the second step is the maximum of f(x) and the minimum of f(x) within the interval a-b. In this example, only $f(-1) = -1$ was derived from the second step. The largest and smallest numbers computed overall are 0 and -1, which represent the minimum and maximum points.

Plot the function $f(x) = x^2 + 2x$ between the interval of $x = 0$ and $x = -2$. Select x values at and near the minimum and maximum points, and solve for f(x).

Values for x are $-3, -2, -1, 0, 1$,
resulting in f(x) values $3, 0, -1, 0, 3$.

Resulting pairs are $(-3,3)$, $(-2, 0)$, $(-1, -1)$, $(0, 0)$, $(1, 3)$.

Graphing the pairs is depicted as:

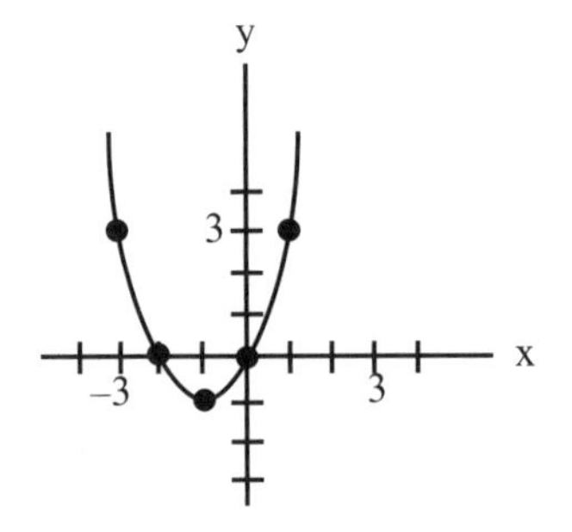

Therefore, because $f'(-1) = 0$ within the interval a-b ($x = -2$ and $x = 0$), at $f(-1)$, the graph of the function $f(x) = x^2 + 2x$ has a minimum.

• By taking the second derivative of a function, it can be determined whether a local extrema is a *minimum* or a *maximum*.

• If point P is in the domain set of the function f and if $f'(P)$ exists and $f'(P) = 0$, then using the second derivative the following is true:

If $f''(P) > 0$ the graph of function f is concave up at P.

If $f''(P) < 0$ the graph of function f is concave down at P.

In other words, if $f'(P)$ exists, and:

If $f'(P) = 0$ and if $f''(P) > 0$ then f has a local minimum at P.

If $f'(P) = 0$ and if $f''(P) < 0$ then f has a local maximum at P.

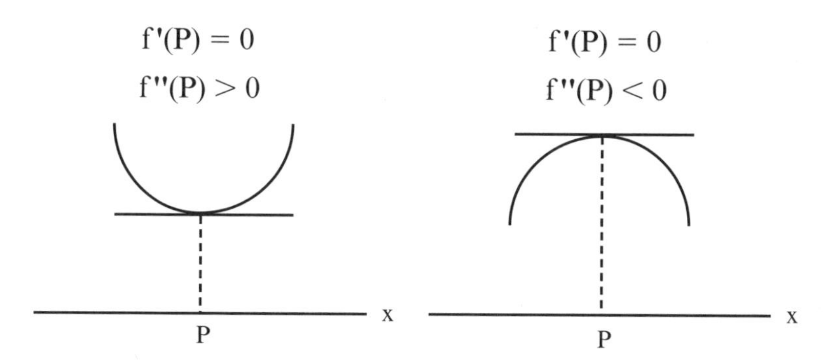

• Remember the previous function $f(x) = x^2 + 2x$, by taking the second derivative determine whether there is a minimum or a maximum at the point where $f'(x) = 0$.

$$f'(x) = \frac{d}{dx}x^2 + \frac{d}{dx}2x = 2x + 2$$

At what value of x does $f'(x) = 0$?

$2x + 2 = 0$

$2x = -2$

$x = -2/2 = -1$

Therefore, at $x = -1$, $f'(x) = 0$.

Taking the second derivative:

$$f''(x) = \frac{d}{dx}2x + \frac{d}{dx}2 = 2 + 0 = 2$$

Using the second derivative rule, because 2 is a positive number, we can predict that the graph of f(x) at $x = -1$ is concave up and is at a minimum. This was depicted in the graph in the previous example.

6.13 Other Rules of Differentiation

• In this section, selected derivatives of the following functions are given. See *Master Math: Calculus* for a more in-depth study.

Function **Derivative**

• $y = u^n$ $$\frac{dy}{dx} = nu^{n-1}\frac{du}{dx}$$

Where n is any positive integer and u is a function of x.

• $y = a^u$ $$\frac{dy}{dx} = a^u \log a \frac{du}{dx}$$

Where u is a function of x.

• $y = u^v$ $$\frac{dy}{dx} = vu^{v-1}\frac{du}{dx} + u^v \log u \frac{dv}{dx}$$

Where u and v are functions of x.

• $y = e^u$ $$\frac{dy}{dx} = e^u \frac{du}{dx}$$

Where u is a function of x.

- $y = \log_a u \qquad \frac{dy}{dx} = \frac{1}{u}\log_a e\frac{du}{dx}$

Where u is a function of x.

- $y = \log u \qquad \frac{dy}{dx} = \frac{1}{u}\frac{du}{dx}$

Where u is a function of x and $u \neq 0$.

- $y = \cos u \qquad \frac{dy}{dx} = -\sin u\frac{du}{dx}$

Where u is a function of x.

- $y = \sin u \qquad \frac{dy}{dx} = \cos u\frac{du}{dx}$

Where u is a function of x.

6.14 An Application of Differentiation: Curvilinear Motion

- Expressions for velocity and acceleration of a particle moving along a curve (curvilinear motion) are more complicated than for a particle moving in a straight line. The equation of the curve can be given in parametric form as $x = f(t)$ and $y = g(t)$, where t represents time. Velocity v is a vector tangent to the curve and has an x and a y component, and is expressed in terms of magnitude (speed) and direction.

Velocity is defined for the x and y components:

The x component of velocity:

$$v_x = \frac{dx}{dt}$$

The y component of velocity:

$$v_y = \frac{dy}{dt}$$

The magnitude (speed) of v:

$$|v| = \sqrt{v_x^2 + v_y^2}$$

The direction of v:

$$\tan \varnothing = \frac{v_y}{v_x} = \frac{dy}{dx}$$

Acceleration is given for the x and y components:

The x component of acceleration:

$$a_x = \frac{dv_x}{dt} = \frac{d^2x}{dt^2}$$

The y component of acceleration:

$$a_y = \frac{dv_y}{dt} = \frac{d^2y}{dt^2}$$

The magnitude of a:

$$|a| = \sqrt{a_x^2 + a_y^2}$$

The direction of a:

$$\tan\varnothing = \frac{a_y}{a_x}$$

The acceleration vector can be expressed in a tangent component and a normal component to the curve.

The tangent component of acceleration:

$$a_T = \frac{v_x a_x + v_y a_y}{|v|}$$

The normal component of acceleration:

$$a_N = \frac{v_x a_y - v_y a_x}{|v|}$$

Chapter 7

Introduction to the Integral

7.1 Definition of the Antiderivative or Indefinite Integral

• This section presents the antiderivative or indefinite integral, the formula for the antiderivative or indefinite integral, the comparison between the antiderivative and the derivative, and the added constant.

• The *antiderivative* or *indefinite integral* is approximately equal to the inverse of the derivative. The antiderivative or indefinite integral of a function f(x) is written:

$$\int f(x)\,dx$$

$\int$ is the integral symbol and f(x) is called the integrand.

• If the derivative of the function f(x) is the function F(x):

$$\frac{df(x)}{dx} = F(x)$$

Then the antiderivative of F(x) is f(x) plus a *constant.*

$$\int F(x)\,dx = f(x) + c$$

c represents an arbitrary constant of integration and dx indicates integrate with respect to x.

• Remember the formula for the derivative:

$$\frac{dx^n}{dx} = nx^{n-1}$$

- The *formula for the antiderivative or indefinite integral* is:

$$\int x^n dx = (\frac{1}{n+1})x^{n+1} + c$$

c represents a constant value and is called the constant of integration.

- A *constant* is added to the integral because the derivative of a constant is zero.

- Evaluate the derivative of the antiderivative formula using the derivative formula:

$$\frac{d}{dx}[(\frac{1}{n+1})x^{n+1} + c] = \frac{d}{dx}[(\frac{1}{n+1})x^{n+1}] + \frac{d}{dx}c$$

$$= (n+1)(\frac{1}{n+1})x^{n+1-1} + 0 = \frac{n+1}{n+1} \times x^{n+0} + 0 = x^n$$

- Because the derivative is the rate of change of some function, it seems likely that several different functions could have the same rate of change. For example, calculate the derivatives of the following three functions:

$$\frac{d}{dx}(2x^2 + 3) = 4x$$

$$\frac{d}{dx}(\sqrt{5} + 2x^2 + 2) = 4x$$

$$\frac{d}{dx}(2x^2 + \pi) = 4x$$

Then, take the integral of 4x to illustrate that each function is different even though they all have the same rate of change (derivative):

$$\int 4x\,dx = (\frac{1}{1+1})4x^{1+1} + c = (\frac{1}{2})4x^2 + c = 2x^2 + c$$

Therefore, in the three functions, c represents

3, $\sqrt{5}$ +2, and π.

7.2 Properties of the Antiderivative or Indefinite Integral

- In this section, the indefinite integral of a function multiplied by a constant, the sum of two functions, a polynomial function, repeated integration, and the integral of a constant are described.

- The indefinite integral of a function multiplied by a number is equal to the number multiplied by the indefinite integral of the function.

$$\int 2\,f(x)\,dx = 2\int f(x)\,d(x)$$

- The indefinite integral of a sum of functions is equal to the sum of the indefinite integrals.

$$\int [f(x) + g(x)]\,dx = \int f(x)\,dx + \int g(x)\,dx$$

• When taking the indefinite integral of a polynomial function, apply the integral formula term-by-term (as with differentiating).

$$\int (x^3 + x^2 + x)\,dx = \int x^3\,dx + \int x^2\,dx + \int x\,dx$$
$$= [(1/4)x^4 + c)] + [(1/3)x^3 + c] + [(1/2)x^2 + c]$$

Combining the c's:

$$= (1/4)x^4 + (1/3)x^3 + (1/2)x^2 + C$$

Where, $C = c + c + c$.

• The derivative is the rate of change of something (as discussed in Chapter 6). For example, the rate of change of distance $x(t)$ is velocity $v(t)$, and the rate of change of velocity $v(t)$ is acceleration $a(t)$.

$$\frac{dx(t)}{dt} = v(t) \quad \text{and} \quad \frac{dv(t)}{dt} = a(t)$$

Conversely, the integral of *acceleration* is velocity, and the integral of *velocity* is distance.

$$\int a(t)\,dt = v(t) + c$$

$$\int v(t)\,dt = x(t) + c$$

• Integration may be repeated any numbers of times. To take multiple integrals of a function, begin with the innermost integral.

For the case of distance = x(t), velocity = v(t), and acceleration = a(t), the double integral of acceleration is:

$$\int\int a(t)\,dt\,dt = \int v(t) + c_1\,dt = x(t) + c_1x + c_2$$

For the case where the $\int$ of f(x) is F(x), the $\int$ of F(x) is g(x), and the $\int$ of g(x) is G(x), then the triple integral of f(x) is:

$$\int\int\int f(x)\,dx\,dx\,dx = \int\int F(x) + c_1\,dx\,dx$$

$$= \int g(x) + c_1x + c_2dx = G(x) + \frac{c_1x^2}{2} + c_2x + c_3$$

- The integral of a constant alone is equal to the constant c multiplied by the variable (in this case x) that the constant is being integrated with respect to (indicated by dx), plus another constant. For example:

$$\int c_1\,dx = c_1x + c_2$$

7.3 Examples of Common Indefinite Integrals

- The following integrals are commonly used in calculus:

$\int 1/x\,dx = \ln|x| + c$ (where ln is the natural logarithm.)

$\int 0\,dx = c$

$\int 2\,dx = 2x + c$

$$\int x^a dx = \frac{x^{a+1}}{a+1} + c \qquad \text{When } a \neq -1.$$

$$\int e^x\, dx = e^x + c$$

$$\int a^x dx = \frac{a^x}{\ln a} + c \qquad \text{When } a > 0 \text{ and } a \neq 1.$$

$$\int \cos x\, dx = \sin x + c$$

$$\int \sin x\, dx = -\cos x + c$$

$$\int \tan x\, dx = \ln|\sec x| + c$$

$$\int \sec^2 x\, dx = \tan x + c$$

$$\int \cot x\, dx = \ln|\sin x| + c$$

$$\int [u(x) + v(x)]\, dx = \int u(x)\, dx + \int v(x)\, dx$$

7.4 Definition and Evaluation of the Definite Integral

- In this section, the *Fundamental Theorem of Calculus,* the integral formula for the definite integral, the definite integral of a function multiplied by a constant, the sum of two functions, improper integrals, and repeated integration are described.

• By the *Fundamental Theorem of Calculus*, if f is a continuous function between points x = a and x = b, and f'(x) is the derivative of f(x), then:

$$\int_a^b f'(x)\,dx = f(b) - f(a)$$

And if F is the antiderivative of f, or F' = f(x), then, the definite integral of f(x) between x = a and x = b is:

$$\int_a^b f(x)\,dx = F(b) - F(a)$$

And the indefinite integral of f(x) is:

$$\int f(x)\,dx = F(x) + c$$

• Definite integrals are evaluated at the values given at the ends of the $\int$ symbol called the *limits of integration.*

Above, F(x) is evaluated at x = b and x = a.

The symbol for evaluated at a and b is $\Big|_a^b$ and it describes subtraction of the function at the top value minus the function at the bottom value:

$$F(x)\Big|_a^b = F(b) - F(a)$$

• Example: Find the area of the function $f(x) = x^2$ between $x = 0$ and $x = 1$ by applying the integral formula and evaluating at the two x boundaries or limits of integration. (See the next section for a discussion on the area under a curve on a graph of a function.)

$$\int_0^1 x^2\,dx = (1/3)\,x^3 \Big|_0^1 = (1/3)(1)^3 - (1/3)(0)^3 = 1/3$$

• The *integral formula for the definite integral* bounded by limits of integration, $x = a$ and $x = b$, is written:

$$\int_a^b \int_a^b x^n dx = \left(\frac{1}{n+1}\right)(x^{n+1})\Big|_a^b = \left(\frac{1}{n+1}\right)(b^{n+1} - a^{n+1})$$

• When the definite integral is evaluated, it is not necessary to add a constant.

• The following two rules for the definite integral of a function multiplied by a constant and the definite integral of a sum of functions that applied to indefinite integrals also apply to definite integrals:

$$\int_a^b Cf(x)\,dx = C\int_a^b f(x)\,dx$$

Where C is any real number.

$$\int_a^b [f(x) + g(x)]\,dx = \int_a^b f(x)\,dx + \int_a^b g(x)\,dx$$

• A definite integral is called an *improper integral* if the integrand is infinite or becomes infinite between its limits, or if one or both of the limits of integration are infinite. An improper integral is discontinuous or diverges at one or more points in the function between the limits of integration.

• Integration may be repeated any number of times. To take multiple integrals of a function, begin with the innermost integral and evaluate it at the limits of integration for the inside integral, then take the integral of the result and evaluate it at the next innermost limits of integration. Repeat this for the number of integrals specified by the number of $\int$ symbols.

• To evaluate the following double integral of the function f(x), take the integral of f(x) and evaluate it at the limits of c and d, then take the integral of the result and evaluate it at the limits of a and b.

$$\int_a^b \int_c^d f(x)\, dx\, dx$$

7.5 The integral and the Area Under the Curve in Graphs of Functions

• One application of the integral is that it can be used to define the area under a curve on a graph of a function. To use the integral to define the area under some region of a curve between points $x = a$ and $x = b$, the curve in this region must be continuous and not extend into a vertical asymptote. Consider the following graph of function f:

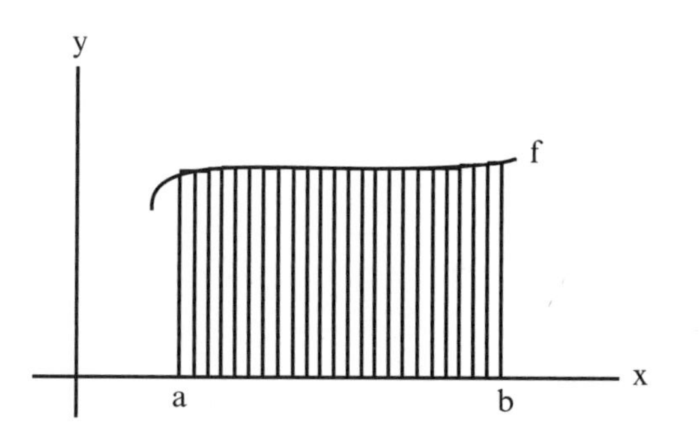

The striped pattern represents the area under the curve of function f. The area under the curve of f between $x = a$ and $x = b$ is given by:

$$\int_a^b f(x)\,dx$$

This is called the *definite integral* of f between $x = a$ and $x = b$.

• In the interval between x = a and x = b on the graph of f(x), the X-axis can be divided into n equal parts of width Δx, such that the Δx segments extend from the X-axis to the f(x) curve so that the area is divided into vertical rectangular strips. (Note that Δ represents some small change in x.)

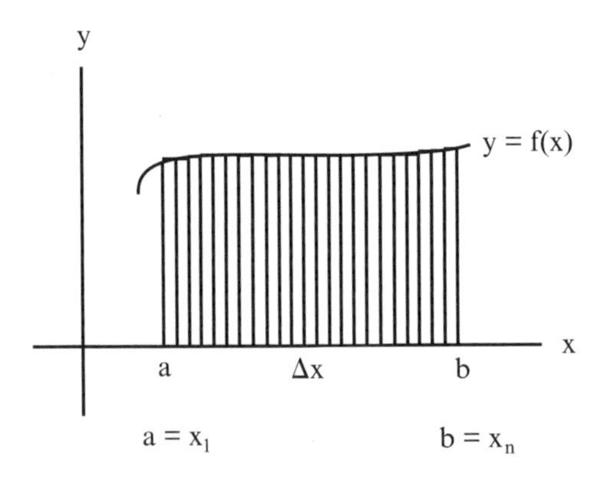

Between x = a and x = b, there are n rectangular strips and each strip is called the *ith strip*. The width of each strip is Δx and the height of each strip is y_i. The area of each strip is width times height, given by:

$$(y_i)(\Delta x) \text{ or } (f(x_i))(\Delta x)$$

An approximation for the total area of f(x) between x = a and x = b is the sum of the areas of the n strips, which can be written:

$$\sum_{i=1}^{n} y_i \Delta x = y_1 \Delta x + y_2 \Delta x + y_3 \Delta x + \ldots + y_1 \Delta x$$

Or equivalently:

$$\sum_{i=1}^{n} f(x_i)\Delta x = f(x_1)\Delta x + f(x_2)\Delta x + f(x_3)\Delta x + \ldots + f(x_i)\Delta x$$

If the width of Δx shrinks and the number of n strips increases, the sum of the strips will represent a better approximation of the actual area. This can be written:

$$\text{Area} = \text{Lim}_{n\to\infty} \sum_{i=1}^{n} y_i \Delta x$$

Also, note that Δx can be equivalently written $\frac{b-a}{n}$.

Therefore, the area approximation can be written:

$$\text{Area} = \text{Lim}_{n\to\infty} \left(\frac{b-a}{n}\right) \sum_{i=1}^{n} y_i$$

The area bound by x = a and x = b is also given by the definite integral and is written:

$$\text{Area} = \int_a^b f(x)\,dx$$

This is approximately equal to:

$$\text{Lim}_{n\to\infty} \sum_{i=1}^{n} f(x_i)\Delta x$$

Area of Functions That Extend Below the X-Axis

- In functions with part of the curve below the X-axis, the area between the X-axis and the curve is negative in value and subtracts from the area above the X-axis.

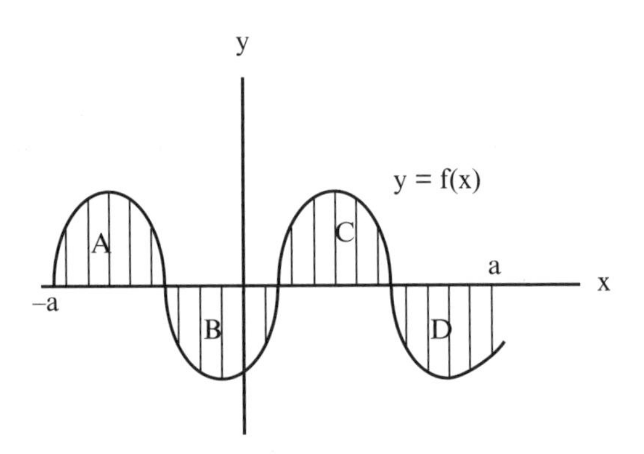

The graph of $y = f(x)$ between $x = -a$ and $x = a$ is given by:

$$\int_{-a}^{a} f(x)\,dx = \text{area A} + \text{area B} + \text{area C} + \text{area D}$$

Area B and area D are negative in value.

- If the area below the curve is equal to the area above the curve, the resulting integral is equal to zero.

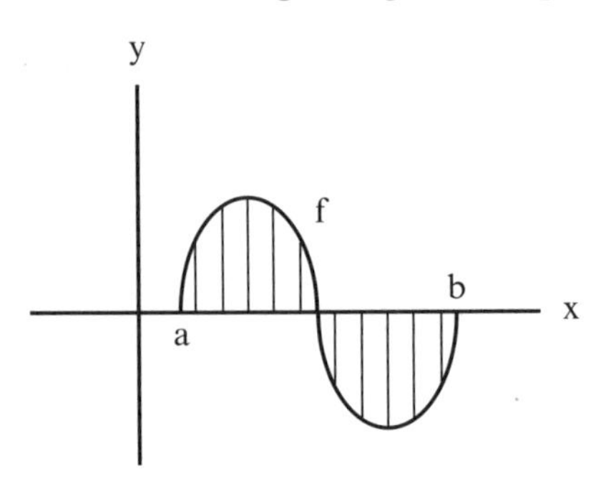

The graph of f between x = a and x = b is given by:

$$\int_a^b f(x)\,dx = \text{positive region} + \text{negative region} = 0$$

The positive region is equal to the negative region.

7.6 Integrals and Volume

• This section will present integrals that define the volume of an object. For a detailed discussion of this subject, *Master Math: Calculus* should be consulted.

• One application of the integral is to define the volume of an object. Volume can be defined by single integral equations, double integral equations, and triple integral equations.

• In the graph of a non-negative continuous function, the area under the curve of function y = f(x) is given by $\int_a^b f(x)\,dx$. If the function is revolved about the X-axis between x = a and x = b, a volume is generated. This volume is called the *volume of revolution.* The area of a vertical cross-section of the volume is πy^2 and the volume of a section that has a thickness of dx is $\pi y^2 dx$. The volume of the volume of revolution between two vertical planes at x = a and x = b is given by:

$$V = \pi \int_a^b [f(x)]^2\,dx \quad \text{or} \quad V = \pi \int_a^b y^2\,dx$$

• In the graph of a non-negative continuous function between x = a and x = b, the volume can be described using the *method of cylindrical shells.* The area bounded by x = a and x = b is revolved about the Y-axis generating a volume. If this volume is divided along the X-axis into n parts each having a thickness of Δx, n vertical cylinders will result. The volume of each shell is obtained by subtracting the volume of a smaller cylinder from the next larger one, $\pi R^2h - \pi r^2h = \pi h(R^2 - r^2)$, where R is the radius of the larger cylinder and r of the smaller, and h is the length along the Y-axis. If the sum of the n shells is taken, as n approaches infinity and the thickness of each shell approaches zero, the volume is described by:

$$V = \int_a^b 2\pi hx\, dx$$

• If the volume of an object is divided into columns in the direction of the Z-axis in an XYZ coordinate system, where the volume of each column is F(x,y) dy dx, then the sum of all the columns gives the total volume for the continuous function F(x,y) and is described by:

$$V = \int_a^b \int_{f1(x)}^{f2(x)} F(x,y)\, dy\, dx$$

• If the volume of an object is divided into cubes in an XYZ coordinate system so that the volume of each cube is dx dy dz, then the sum of all the cubes gives the total volume and is described by:

$$V = \int_{a}^{b} \int_{f1(x)}^{f2(x)} \int_{F1(x)}^{F2(x)} dz\, dy\, dx$$

7.7 Even Functions, Odd Functions, and Symmetry

• By determining whether a function is even or odd, it is often possible to simplify the integral of the function to a more manageable form and solve using symmetry.

• A function is *even* if $f(x) = f(-x)$ between $x = -a$ and $x = a$. An example of a graph of an *even function* is:

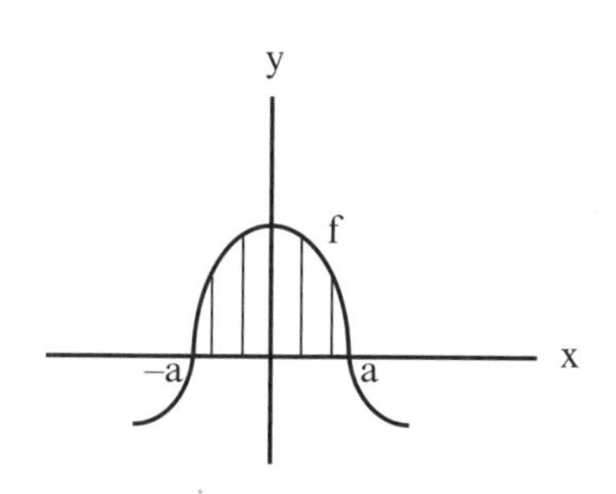

From the graph, it is clear by symmetry that the section on the left of the Y-axis, between $x = -a$ and $x = 0$, is equivalent to the section on the right of the Y-axis, between $x = 0$ and $x = a$.

The integral for this even function can be written:

$$\int_{-a}^{a} f(x)\,dx = 2\int_{-a}^{0} f(x)\,dx = 2\int_{0}^{a} f(x)\,dx$$

In an even function, the area for negative values of x is equal to the area for positive values of x.

- Examples of even functions are:

  $f(x) = c$

  $f(x) = x^2$

  $f(x) = x^4$

  $f(x) = x^{2n}$

  Note: $f((-x)^2) = (-x)(-x) = x^2$.

- A function is *odd* if $f(x) = -f(-x)$ between $x = -a$ and $x = a$. An example of a graph of an *odd function* is:

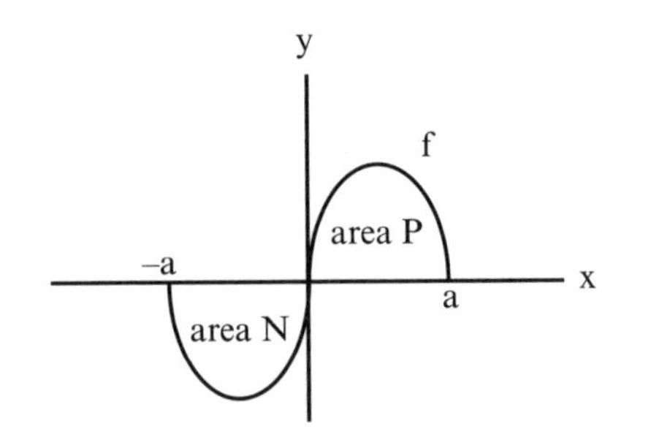

From the graph, it is clear by symmetry that the section on the left of the Y-axis, between $x = -a$ and $x = 0$, is equivalent but opposite to the section on the right of the Y-axis, between $x = 0$ and $x = a$.

The integral for this odd function can be written:

$$\int_{-a}^{a} f(x)\,dx = \int_{-a}^{0} f(x)\,dx + \int_{0}^{a} f(x)\,dx = \text{areaP} + \text{areaN} = 0$$

Note that $\int_{-a}^{0} f(x)\,dx = -\int_{0}^{a} f(x)\,dx$.

The area for negative values of x is equal but opposite to the area for positive values of x, and the two areas subtract and cancel each other out, resulting in an integral equivalent to zero.

- Examples of odd functions are:

  $f(x) = x$

  $f(x) = x^3$

  $f(x) = x^5$

  $f(x) = x^{2n+1}$

  Note: $f((-x)^3) = (-x)(-x)(-x) = (-x)^3$

7.8 Properties of the Definite Integral

- The following are properties of definite integrals:

$$\int_a^b f(x)\,dx = -\int_b^a f(x)\,dx$$

$$\int_a^b Cf(x)\,dx = C\int_a^b f(x)\,dx \quad \text{Where C is any constant.}$$

$$\int_a^b [f(x) + g(x)]\,dx = \int_a^b f(x)\,dx + \int_a^b g(x)\,dx$$

$$\int_a^a f(x)\,dx = 0$$

$$\int_a^c f(x)\,dx = \int_a^b f(x)\,dx + \int_b^c f(x)\,dx \quad \text{When } a < b < c.$$

7.9 Methods for Evaluating Complex Integrals: Integration by Parts, Substitution, and Tables

- To evaluate complicated integrals, methods that go beyond simply applying the integral formula are often required. Some of the most common methods are integration by parts, substitution, and looking up integrals in integral tables.

- To evaluate certain complicated integrals, the method of *integration by parts* can be applied. To use this method, the integral must exist in the form or be arranged to fit the form of the following formula:

$$\int f(x)\,g'(x)\,dx = f(x)\,g(x) - \int f'(x)\,g(x)\,dx$$

Using other notation, where $u = f(x)$ and $v = g(x)$, the formula is written:

$$\int u\,dv = uv - \int v\,du$$

• In the following example, integration by parts can be used to solve the integral:

$$\int x \cos x\,dx$$

To arrange this integral in a form that will fit the integration by parts formula, make the following substitutions:

$$u = x$$

$$du = dx$$

$$dv = \cos x\,dx$$

From the substitution for dv, $v = \int \cos x\,dx = \sin x$.

Using the integration by parts formula:

$$\int u\,dv = uv - \int v\,du$$

The integral becomes:

$$\int x \cos x\,dx = x \sin x - \int \sin x\,dx$$

Because $\int \sin x\,dx = -\cos x$, the evaluated integral becomes:

$$\int x \cos x\,dx = x \sin x + \cos x + c$$

• *Substitution* of variables is used to translate a complicated integral into a more manageable form so that the integral can be solved using the integral formula or integral tables, then the integral is translated back to its original variables. For example, given the function:

$$f(x)\,dx = \frac{x+1}{x^2 + 2x + 10}$$

The integral is $\int f(x)\,dx = \int \left(\frac{x+1}{x^2 + 2x + 10} \right) dx$

First, simplify the integral by substituting:

$$u = (x^2 + 2x + 10)$$

Where the derivative of u is:

$$du = (2x + 2)\,dx = 2(x + 1)\,dx$$

Rearranging to isolate dx gives:

$$dx = du / 2(x + 1)$$

Next, substitute u and du into the integral so that $(x^2 + 2x + 10) = u$ and $dx = du / 2(x + 1)$, then evaluate the integral.

$$\int \frac{(x+1)\;du}{(u)\,(2)\,(x+1)} = \int \frac{1}{2u}\,du = \frac{1}{2}\int \frac{1}{u}\,du = \frac{1}{2}\ln |u| + c$$

(Remember, $\int 1/x\,dx = \ln|x| + c$.)

Finally, substitute the original expressions back into the evaluated integral, where $u = (x^2 + 2x + 10)$.

$$= (1/2) \ln |(x^2 + 2x + 10)| + c$$

Therefore,

$$\int (x + 1)/(x^2 + 2x + 10)\, dx = (1/2) \ln |(x^2 + 2x + 10)| + c$$

Integral Tables

- There are many advanced techniques for solving complex integrals that can be found in calculus textbooks and handbooks of mathematical functions. Integral tables are useful for solving integrals in forms that do not allow easy application of integration techniques. Integral tables are found in mathematical handbooks, calculus books, on-line integral tables and resources, and the *CRC Handbook of Chemistry and Physics.*

- Integral tables contain solved integrals in various forms so that an unknown integral can be matched to, or translated into, the form in the integral table that is identical or most similar to it. If the unknown integral is not identical to a form in the table, a transformation of the integral must be made using substitution, for example, substitute y for ax. Specific substitutions are suggested within integral tables for certain integrals. In general, when making substitutions, it is important to make a substitution of the dx terms, to express the limits of the definite integrals in the new dependent variable, and to perform reverse substitution to obtain the answer in terms of the original independent variable for indefinite integrals.

It may be helpful to peruse some integral tables and on-line resources to become familiar with the integrals and substitution suggestions, and to read the introductory discussions at the beginning of the tables.

Index

D

E

F

G

S

T

U–V

X

Y

Z

Notes

Notes

Notes

Notes

Notes

Notes

Notes

Notes